Code of Safety for
Fishermen and Fishing Vessels
2005

Part B
Safety and Health Requirements for the
Construction and Equipment of Fishing Vessels

INTERNATIONAL
MARITIME
ORGANIZATION
London, 2006

Published by the
INTERNATIONAL MARITIME ORGANIZATION
4 Albert Embankment, London SE1 7SR

Printed and bound in the United Kingdom by William Clowes Ltd, Beccles, Suffolk

2 4 6 8 10 9 7 5 3 1

First edition, 1975
2nd edition, 2006

ISBN-13: 978-92-801-4209-9
ISBN-10: 92-801-4209-7

IMO PUBLICATION
Sales number: IA755E

Copyright © International Maritime Organization 2006

All rights reserved.
No part of this publication may be reproduced,
stored in a retrieval system or transmitted in any form
or by any means without prior permission in writing
from the International Maritime Organization.

Preface

1   The Code of Safety for Fishermen and Fishing Vessels originated from a resolution adopted by the International Labour Organization (ILO) in 1962. Subsequent to that resolution, the Food and Agriculture Organization (FAO), ILO and the International Maritime Organization (IMO) entered into an agreement to co-operate, within their respective fields of experience, to elaborate the Code. The agreement acknowledged that the respective areas of competence are:

- FAO – fisheries in general;
- ILO – labour in the fishing industry; and
- IMO – safety of life, vessels and equipment at sea.

The Code was elaborated in two parts:

.1   part A to be addressed to skippers and crews, containing operational and occupational requirements; and

.2   part B to be addressed to shipbuilders and owners, containing requirements for the construction and equipment for fishing vessels.

2   Part A of the Code was adopted by the first session of the Joint FAO/ILO/IMO Meeting of Consultants on Safety on Board Fishing Vessels which was held at ILO Headquarters in Geneva in September 1968.

3   Later amendments to part A were approved by the Maritime Safety Committee (MSC) of IMO, at its thirtieth session in the spring of 1973. At the same session, the Committee approved the final text of part B which was endorsed by the FAO Council at its 64th session (autumn 1974) and also endorsed by the Governing Body of the ILO at its 195th session (February 1975).

4   In 1977, an International Conference on the Safety of Fishing Vessels adopted the Torremolinos International Convention for the Safety of Fishing Vessels, 1977 which, for a number of reasons, did not enter into force. Consequently, a further International Conference was convened, also in Torremolinos, Spain, that adopted the Torremolinos Protocol of 1993 relating to the Torremolinos International Convention for the Safety of Fishing Vessels, 1977, hereinafter referred to as the Protocol.

5   The Conference also adopted, *inter alia*, resolution 4 in which it is noted that the Protocol does not contain specific requirements for certain safety equipment for fishing vessels of less than 45 m in length, such as life-saving appliances. Consequently, it urged all States, in view of the

inherent risks involved in the operation of fishing vessels, to consider the requirements for safety equipment when deciding, in accordance with article 3(4) of the Protocol, which regulations they should apply, wholly or in part, to fishing vessels of 24 m in length and over but less than the applicable length criteria of the chapter in question.

6 It was also noted that initiatives had been taken by certain States to develop uniform regional standards as called for in article 3(5) of the Protocol to ensure that the safety of fishing vessels covered by article 3(4) thereof is maintained at an acceptable level by determining which regulations, contained in the annex to the Protocol should apply, wholly or in part, to such vessels.

7 In its review of regional standards so developed, the MSC of IMO noted that they had been examined by the Sub-Committee on Stability and Load Lines and on Fishing Vessels Safety (SLF) with a view to the desirability of developing a template for other countries or regions (see documents MSC 68/INF.10 and MSC 70/INF.24). It was also noted that, in the examination process, the provisions of these regional standards had provided valuable information in relation to the revision of part B of the Code.

8 The MSC accepted that for certain sizes of vessels, the minimum standards contained in the Protocol should be applied and considered that it would be appropriate to refer to such provisions of the Protocol, where relevant, in the revised text of part B of the Code. It was also accepted by the Committee that any such references must stress the voluntary nature of the Code and substitute the mandatory terms "shall" and "will" with the word "should".

9 The MSC acknowledged that there had been significant developments in relation to the management of fisheries that contained principles in support of the safety of fishermen and fishing vessels. It noted in particular:

.1 the agreement for the Interpretation of the Provisions of the United Nations Convention on the Law of the Sea of 10 December 1982 Relating to the Conservation and Management of Straddling Fish Stocks and Highly Migratory Fish Stocks of 1995; and

.2 the Code of Conduct for Responsible Fisheries adopted by the Conference of FAO in 1995.

10 The MSC recognized that the safety at sea aspects contained within these instruments could be relevant in relation to the revision of part B, in particular:

.1 the arrangements for the monitoring, control and surveillance of fishing vessels including recommendations for the reporting of the position of a fishing vessel at sea;

Preface

    .2   the marking of fishing vessels in accordance with uniform and internationally recognized systems such as the FAO Standard Specifications for the Marking and Identification of Fishing Vessels*; and

    .3   the integration of fishing vessels into search and rescue systems.

11   In entrusting the revision of the Code to the SLF Sub-Committee, the MSC recommended that the recent developments in fishing vessel design and fishing operations should be taken into consideration. The MSC also entrusted the SLF Sub-Committee to revise the Voluntary Guidelines for the Design, Construction and Equipment of Small Fishing Vessels, that had been approved by MSC in 1979, which addresses vessels of 12 m in length and over but less than 24 m in length. In this regard, IMO was requested to invite FAO and ILO to participate in the revision. Both Organizations responded positively to the subsequent invitation.

12   The SLF Sub-Committee established a correspondence group to facilitate the revision of the Code and the Voluntary Guidelines and, following clearance by the relevant sub-committees of the IMO, the revised text was submitted to the MSC at its seventy-ninth session (1 to 10 December 2004) at which it was approved. At the twenty-sixth session of the Committee on Fisheries in March 2005, FAO welcomed the revised Code and Voluntary Guidelines and recommended the early publication by IMO of these documents. The Governing Body of ILO approved the revised texts at its 293rd session in June 2005.

13   Concerning the procedures for future amendments to the Code and the Voluntary Guidelines, the MSC considered that any amendments should be effected as expeditiously as possible. It was agreed that non-controversial amendments should be approved by correspondence but joint meetings of experts might be necessary for other amendments for which no ready agreement by correspondence could be reached.

14   Recognizing that the majority of items covered by the Code and the Voluntary Guidelines are within the scope of IMO and noting the different working procedures within the three Organizations and also that the SLF Sub-Committee holds regular meetings, it was agreed that:

    .1   IMO should act as a focal point for co-ordinating proposed amendments to the Code and in particular the IMO Secretariat should undertake to receive any proposed amendments, to distribute them to the Organizations and to collate their respective comments;

    .2   any future joint FAO/ILO/IMO meeting should be held, whenever possible, in conjunction with a meeting of the SLF Sub-Committee; and

---

* Refer to the FAO Technical Guidelines for Responsible Fisheries – No 1 Fishing Operations. (ISBN 92-5-103914-3) and MSC/Circ.572.

.3 any proposed amendments should always be subject to the final approval of the appropriate bodies of the three Organizations.

Contents

		Page
Chapter I	**General provisions**	1
Chapter II	**Construction, watertight integrity and equipment**	7
Chapter III	**Stability and associated seaworthiness**	16
Chapter IV	**Machinery and electrical installations and periodically unattended machinery spaces**	
Part A	General.	24
Part B	Machinery installations	28
Part C	Electrical installations	39
Part D	Periodically unattended machinery spaces.	44
Chapter V	**Fire protection, fire detection, fire extinction and fire fighting**	
Part A	General fire protection provisions	48
Part B	Fire safety measures in vessels of a length of 60 m and over.	51
Part C	Fire safety measures in vessels of 45 m in length and over but less than 60 m	76
Part D	Fire safety measures in vessels of 24 m in length and over but less than 45 m	87
Chapter VI	**Protection of the crew**.	99
Chapter VII	**Life-saving appliances and arrangements**	
Part A	General.	111
Part B	Vessel requirements	113
Part C	Life-saving appliance requirements.	119

Code of Safety for Fishermen and Fishing Vessels – Part B

		Page
Chapter VIII	**Emergency procedures, musters and drills**	120
Chapter IX	**Radiocommunications**	
Part A	General	127
Part B	Ship requirements	130
Chapter X	**Shipborne navigational equipment and arrangements**	139
Chapter XI	**Crew accommodation**	145
Annex I	**Illustration of terms used in the definitions**	153
Annex II	**Recommended practice for anchor and mooring equipment**	157
Annex III	**Recommended practice on portable fish-hold divisions**	162
Annex IV	**Recommended practice for ammonia refrigeration systems in manned spaces**	166
Annex V	**Recommendations for testing lifejackets and lifebuoys**	169
Annex VI	**Recommended standards for pilot ladders**	184
Annex VII	**Annotated list of pertinent publications**	188
Information Note: Fisheries management measures		193
Index		197

Chapter I

General provisions

1.1 Purpose and scope

1.1.1 The purpose of this part of the Code is to provide information on the design, construction, and equipment of fishing vessels with a view to promoting the safety of fishing vessels and safety and health of the crew. The Code is not a substitute for national laws and regulations nor is it a substitute for the provisions of international instruments in relation to safety of fishing vessels and crew although it may serve as a guide to those concerned with framing such national laws and regulations.

1.1.2 The Code is voluntary. It is wider in scope than the Torremolinos Protocol and only the minimum requirements to ensure the safety of fishing vessels and safety and health of the crew are given in this part of the Code for fishing vessels of 24 m in length and above. Each Competent Authority should take every possible measure to promote the safety of the vessels concerned.

1.1.3 Certain sections of this part of the Code make reference to the minimum standards set out in the provisions of the Torremolinos International Convention for the Safety of Fishing Vessels, 1977, as modified by the Torremolinos Protocol of 1993 relating thereto. For the purpose of this part of the Code, these are considered to be the minimum standards acceptable in relation to the classes of vessels, as prescribed in the Protocol, to which they should be applied.

1.1.4 Regional uniform standards or guidelines that have been submitted to IMO as provided for under article 3, paragraphs (4) and (5) of the Protocol for fishing vessels registered and operating in such regions, prevail over chapters IV, V, VII and IX of this part of the Code. For all other fishing vessels of 24 m in length and over but less than 45 m in length, that are registered in such regions but operate, or are intended for operation outside the region, the whole of this part of the Code would serve as a guide.

1.1.5 Unless otherwise stated, the provisions of this part of the Code are intended to apply to new decked fishing vessels of 24 m in length and above. However, even where not otherwise stated, the Competent Authority should also apply these provisions, as far as reasonable and practicable, to existing decked fishing vessels.

1.1.6 The provisions of this part of the Code do not apply to fishing vessels for sport or recreation or to processing vessels.

1.1.7 Where operating experience has clearly shown that departure from the provisions of this part of the Code is justified, or in applying this part of the Code to any other equivalent area of operation for any vessel covered by this part of the Code, the Competent Authority may permit adequate alterations or substitutions thereof.

1.2 Definitions

For the purpose of this part of the Code, unless expressly provided otherwise, the following definitions apply.

1.2.1 *Amidships* is the mid-length of L.

1.2.2 *Approved* means approved by the Competent Authority.

1.2.3 *Baseline* is the horizontal line intersecting at amidships the keel line.

1.2.4 *Bow height*, defined as the vertical distance at the forward perpendicular between the waterline corresponding to the maximum permissible operating draught and the designed trim and the top of the exposed deck at side.

1.2.5 *Breadth (B)** is the maximum breadth of the vessel, measured amidships to the moulded line of the frame in a vessel with a metal shell and to the outer surface of the hull in a vessel with a shell of any other material.

1.2.6 *Collision bulkhead* is a watertight bulkhead up to the working deck in the forepart of the vessel which meets the following conditions:

- .1 The bulkhead should be located at a distance from the forward perpendicular:
    - .1.1 not less than $0.05L$ and not more than $0.08L$ for vessels of 45 m and over;
    - .1.2 not less than $0.05L$ and not more than $0.05L$ plus 1.35 m for vessels less than 45 m in length except as may be allowed by the Competent Authority; and
    - .1.3 in no case less than 2 m.
- .2 Where any part of the underwater body extends forward of the forward perpendicular, e.g. a bulbous bow, the distance stipulated in 1.2.6.1 should be measured from a point at mid-length of the extension forward of the forward perpendicular or from a point $0.015L$ forward of the forward perpendicular, whichever is less.

---

* Definitions are illustrated in annex I to this part of the Code.

.3 The bulkhead may have steps or recesses provided they are within the limits prescribed in 1.2.6.1.

1.2.7 *Competent Authority* is the Government of the State whose flag the vessel is entitled to fly.

1.2.8 *Convention* means the International Convention for the Safety of Life at Sea, 1974, as amended.

1.2.9 *Crew* means the skipper and all persons employed or engaged in any capacity on board a vessel on the business of that vessel.

1.2.10 *Deepest operating waterline* is the waterline related to the maximum permissible operating draught.

1.2.11 The *depth (D)* is the moulded depth amidships.

1.2.12 *Enclosed superstructure* is a superstructure with:

.1 enclosing bulkheads of efficient construction;

.2 access openings, if any, in those bulkheads fitted with permanently attached weathertight doors of a strength equivalent to the unpierced structure which can be operated from each side; and

.3 other openings in sides or ends of the superstructure fitted with efficient weathertight means of closing.

A raised quarter-deck is regarded as a superstructure.

A bridge or poop should not be regarded as enclosed unless access is provided for the crew to reach machinery and other working spaces inside those superstructures by alternative means which are available at all times when bulkhead openings are closed.

1.2.13 *Existing vessel* is a vessel which is not a new vessel.

1.2.14 *Fishing vessel* – in the following referred to as "vessel" – is a vessel used commercially for catching fish, whales, seals, walrus, or other living resources of the sea.

1.2.15 The *forward and after perpendiculars* should be taken at the forward and after ends of the length (L). The forward perpendicular should be coincident with the foreside of the stem on the waterline on which the length is measured.

1.2.16 *Freeboard* (f_{min}) is the actual minimum freeboard and is the distance from the underside of the working deck at the side to a waterline, measured perpendicularly to the waterline, plus the minimum thickness of decking. When the working deck is stepped, the lowest line of the deck and the continuation of that line parallel to the upper part of the deck should be taken as the working deck.

Chapter I

1.2.17 *Height of a superstructure or other erection* is the least vertical distance measured at side from the top of the deck beams of a superstructure or an erection to the top of the working deck beams.

1.2.18 *Keel line** is the line parallel to the slope of keel passing amidships through:

    .1 the top of the keel or line of intersection of the inside of shell plating with the keel where a bar keel extends above that line of a vessel with a metal shell;

    .2 the rabbet lower line of the keel of a vessel with a shell of wood or a composite material; or

    .3 the intersection of a fair extension of the outside of the shell contour at the bottom with the centreline of a vessel with a shell of material other than wood and metal.

1.2.19 *Least depth** is the depth measured from the keel line to the top of the working deck beam at side at the point where a parallel to the keel line is tangent to the deck line. Where the working deck is stepped and the raised part of the deck extends over the point at which the least depth is to be determined, the least depth should be measured to a line of reference extending from the lower part of the deck along a line parallel with the raised part.

1.2.20 *Length (L)** should be taken as 96% of the total length on the waterline at 85% of the least depth measured from the keel line, or as the length from the foreside of the stem to the axis of the rudder stock on that waterline, if that be greater. In vessels designed with rake of keel, the waterline on which this length is measured should be parallel to the designed waterline.

1.2.21 *Midship section* is that section of the hull defined by the intersection of the moulded surface of the hull with a vertical plane perpendicular to the water and centreline planes passing through amidships.

1.2.22 The *moulded depth* is the vertical distance measured from the keel line to the top of the working deck beam at side. In vessels having rounded gunwales, the moulded depth should be measured to the point of intersection of the moulded lines of deck and side shell plating, the lines extending as though the gunwale were of angular design. Where the working deck is stepped and the raised part of the deck extends over the point at which the moulded depth is to be determined, the moulded depth is measured to a line of reference extending from the lower part of the deck along a line parallel with the raised part.

1.2.23 *New vessel* is a vessel the keel of which is laid, or which is at a similar stage of construction, on or after the date of adoption of the present revision to this part of the Code.

---

* Definitions are illustrated in annex I to this part of the Code.

1.2.24 *Organization* means the International Maritime Organization.

1.2.25 *Place of shelter* is any naturally or artificially protected area easily accessible to the vessel and which can be used for sheltering the vessel in circumstances which are unfavourable to its safety.

1.2.26 *Processing vessel* is a vessel used exclusively for processing fish and other living resources of the sea.

1.2.27 *Protocol* means the Torremolinos International Convention for the Safety of Fishing Vessels, 1977, as modified by the Torremolinos Protocol of 1993 relating thereto.

1.2.28 *Skipper* means the person having command of a fishing vessel.

1.2.29 *Superstructure* is the decked structure on the working deck extending from side to side of the vessel or with the side plating not being inboard of the shell plating more than $0.04B$.

1.2.30 *Superstructure deck* is that complete or partial deck or the top of a superstructure, deckhouse or other erections situated at a height of not less than 1.8 m above the working deck. Where this height is less than 1.8 m, the top of such deckhouses or other erections should be treated in the same way as the working deck.

1.2.31 *Watertight* means capable of preventing the passage of water through the structure in any direction under a head of water for which the surrounding structure is designed.

1.2.32 *Weather deck* is the uppermost deck exposed to weather and sea. Where the deck is not continuous, the uppermost deck at the point in question should be taken as the weather deck.

1.2.33 *Weathertight* means that in any sea conditions water will not penetrate into the vessel.

1.2.34 *Working deck* is generally the lowest complete deck above the deepest operating waterline from which fishing is undertaken. In vessels fitted with two or more complete decks, the Competent Authority may accept a lower deck as a working deck provided that that deck is situated above the deepest operating waterline.

In this part of the Code measurements are given in the metric system, using the following abbreviations:

m	–	metre
cm	–	centimetre
mm	–	millimetre
t	–	tonne (1000 kg)
kg	–	kilogram
m·t	–	metre-tonne
°C	–	degree centigrade

s   —   second
N   —   newton
kW —   kilowatt

1.3 Surveys

1.3.1 The hull, machinery, equipment, and radio installations should be surveyed on completion and thereafter in such manner and at such intervals as the Competent Authority may consider necessary in order to ensure that their condition is in all respects satisfactory. The Competent Authority may, however, entrust the inspections and surveys either to surveyors nominated for the purpose or to organizations recognized by it. The surveys should be such as to ensure that the arrangements, material, and scantlings of the structure, boilers and other pressure vessels and their appurtenances, main and auxiliary machinery, electrical installations as well as crew accommodation and other equipment are in all respects satisfactory for the service for which the vessel is intended.

1.3.2 After any survey has been completed, no change should be made in the structural arrangements, machinery, equipment, etc., covered by the survey, without the sanction of the Competent Authority.

1.3.3 A fishing vessel should carry on board documentation relating to the safety of the vessel issued by the Competent Authority.

1.3.4 Documentation relating to the safety of the vessel should cease to be valid upon transfer of the vessel to the flag of another State. New safety documentation should only be issued when the Competent Authority is fully satisfied that the vessel is in compliance with the requirements of the relevant provisions.

1.4 Equivalents

Where the present provisions require that a particular fitting, material, appliance or apparatus, or type thereof, should be fitted or carried in a vessel, or that any particular provision should be made, the Competent Authority may allow any other fitting, material, appliance or apparatus, or type thereof, to be fitted or carried, or any other provision to be made in that vessel, if it is satisfied by trial thereof or otherwise that such fitting, material, appliance or apparatus, or type thereof, or provision, is at least as effective as that required by the present provisions. Documentation approved by the Competent Authority to the effect that the alternative structures, measures and appliances meet national regulations should be available on board a vessel.

Chapter II
Construction, watertight integrity and equipment

2.1 Construction

2.1.1 Strength and construction of hull, superstructures, deckhouses, machinery casings, companionways and any other structures and vessel's equipment should be sufficient to withstand all foreseeable conditions of the intended service and should be to the satisfaction of the Competent Authority.

2.1.2 The hull of vessels intended for operation in ice should be strengthened in accordance with the anticipated conditions of navigation and area of operation.

2.1.3 Bulkheads, closing devices and closures of openings in these bulkheads, as well as methods for their testing, should be in accordance with the requirements of the Competent Authority. Vessels constructed of material other than wood should be fitted with a collision bulkhead and at least with watertight bulkheads bounding the main machinery space. Such bulkheads should be extended up to the working deck. In vessels constructed of wood, such bulkheads, which as far as practicable should be watertight, should also be fitted.

2.1.4 Watertight doors fitted in watertight bulkheads should be capable of being opened and closed locally at the door on either side and preferably also from above the working deck. Means of operating doors should be clearly marked, and should indicate whether doors are open or closed. Pipes piercing the collision bulkhead should be fitted with suitable valves operable from above the working deck and the valve chest should be secured at the collision bulkhead inside the forepeak. No door, manhole, ventilation duct or any other opening should be fitted in the collision bulkhead below the working deck.

2.1.5 Where a long forward superstructure is fitted, the collision bulkhead should be extended weathertight to the deck next above the working deck. The extension need not be fitted directly over the bulkhead below provided it is located within the limits given in 1.2.6 and the part of the deck which forms the step is made effectively weathertight.

2.1.6 The number of openings in the collision bulkhead above the working deck should be reduced to the minimum compatible with the design and normal operation of the vessel. Such openings should be capable of being closed weathertight.

2.1.7 In vessels of 75 m in length and over, a watertight double bottom should be fitted, as far as practicable, between the collision bulkhead and the afterpeak bulkhead.

2.2 Watertight doors

2.2.1 The number of openings in watertight bulkheads, as required by 2.1.3, should be reduced to the minimum compatible with the general arrangements and operational needs of the vessel; openings should be fitted with watertight closing appliances to the satisfaction of the Competent Authority. Watertight doors should be of an equivalent strength to the adjacent unpierced structure.

2.2.2 In vessels of less than 45 m in length, such doors may be of the hinged type, which should be capable of being operated locally from each side of the door and should normally be kept closed at sea. A notice should be attached to the door on each side to state that the door should be kept closed at sea.

2.2.3 In vessels of 45 m in length and over, watertight doors should be of the sliding type in:

.1 spaces where it is intended to open them at sea and if located with their sills below the deepest operating waterline, unless the Competent Authority considers it to be impracticable or unnecessary, taking into account the type and operation of the vessels; and

.2 the lower part of a machinery space where there is access from it to a shaft tunnel.

Otherwise watertight doors may be of the hinged type.

2.2.4 Sliding watertight doors should be capable of being operated when the vessel is listed up to 15° either way.

2.2.5 Sliding watertight doors, whether manually operated or otherwise, should be capable of being operated locally from each side of the door; in vessels of 45 m in length and over, these doors should also be capable of being operated by remote control from an accessible position above the working deck except when the doors are fitted in crew accommodation spaces.

2.2.6 Means should be provided at remote operating positions to indicate when a sliding door is open or closed.

2.3 Hull integrity

2.3.1 External openings should be capable of being closed so as to prevent water from entering the vessel. Deck openings which may be open during fishing operations should normally be arranged near to the vessel's centreline. However, the Competent Authority may approve different arrangements if satisfied that the safety of the vessel will not be impaired.

2.3.2 Fish flaps on stern trawlers should be power-operated and capable of being controlled from any position which provides an unobstructed view of the operation of the flaps.

2.4 Weathertight doors

2.4.1 All access openings in bulkheads of enclosed superstructures and other outer structures, through which water could enter and endanger the vessel, should be fitted with doors permanently attached to the bulkhead, framed and stiffened so that the whole structure is of equivalent strength to the unpierced structure, and weathertight when closed. The means for securing these doors weathertight should consist of gaskets and clamping devices or other equivalent means and should be permanently attached to the bulkhead or to the doors themselves, and should be so arranged that they can be operated from each side of the bulkhead. The Competent Authority may, without prejudice to the safety of the crew, permit the doors to be opened from one side only for freezer rooms, provided that a suitable alarm device is fitted to prevent persons being trapped in those rooms.

2.4.2 The height above deck of sills in those doorways, in companionways, erections and machinery casings which give direct access to parts of the deck exposed to the weather and sea should be at least 600 mm on the working deck and at least 300 mm on the superstructure deck. Where operating experience has shown justification and on approval by the Competent Authority, these heights, except in the doorways giving direct access to machinery spaces, may be reduced to not less than 380 mm and 150 mm respectively.

2.5 Hatchways closed by wood covers

2.5.1 The height above deck of hatchway coamings should be at least 600 mm on exposed parts of the working deck and at least 300 mm on the superstructure deck.

2.5.2 The finished thickness of wood hatchway covers should include an allowance for abrasion due to rough handling. In any case, the finished thickness of these covers should be at least 4 mm for each 100 mm of unsupported span, subject to a minimum of 40 mm, and the width of their bearing surfaces should be at least 65 mm.

2.5.3 Arrangements for securing wood hatchway covers weathertight should be provided to the satisfaction of the Competent Authority.

2.6 Hatchways closed by covers other than wood

2.6.1 The height above deck of hatchway coamings should be as specified in 2.5.1. Where operating experience has shown justification and on the approval by the Competent Authority, the height of these coamings may be reduced, or the coamings omitted entirely, provided that the safety of vessels is not thereby impaired. In this case, the hatchway openings should be kept as small as practicable and the covers be permanently attached by hinges or equivalent means and be capable of being rapidly closed and battened down, or by equally effective arrangements to the satisfaction of the Competent Authority.

2.6.2 For the purpose of strength calculations, it should be assumed that hatchway covers are subjected to the weight of cargo intended to be carried on them or to the following static loads, whichever is the greater:

.1   10 kN/m^2 for vessels of 24 m in length;

.2   17 kN/m^2 for vessels of 100 m in length and over.

For intermediate lengths, the load values should be determined by linear interpolation. The Competent Authority may reduce the loads to not less than 75% of the above values for covers to hatchways situated on the superstructure deck in a position abaft a point located 0.25L of the length of the vessel measured from the forward perpendicular.

2.6.3 Where covers are made of mild steel, the maximum stress calculated according to 2.6.2 multiplied by 4.25 should not exceed the minimum ultimate strength of the material. Under these loads, the deflections should not be more than 0.0028 times the span.

2.6.4 Covers made of materials other than mild steel should be at least of equivalent strength to those made of mild steel, and their construction should be of sufficient stiffness ensuring weathertightness under the loads specified in 2.6.2.

2.6.5 Covers should be fitted with clamping devices and gaskets sufficient to ensure weathertightness, or other equivalent arrangements to the satisfaction of the Competent Authority.

2.7 Machinery space openings

2.7.1 Machinery space openings should be framed and enclosed by casings of a strength equivalent to the adjacent superstructure. External access openings therein should be fitted with doors complying with the provisions of 2.4.

2.7.2 Openings other than access openings should be fitted with covers of equivalent strength to the unpierced structure, permanently attached thereto and capable of being closed weathertight.

2.8 Other deck openings

2.8.1 Where it is essential for fishing operations, flush deck scuttles of the screw, bayonet or equivalent type and manholes may be fitted provided these are capable of being closed watertight and such devices should be permanently attached to the structure. Having regard to the size and disposition of the openings and the design of the closing devices, metal-to-metal closures may be fitted if the Competent Authority is satisfied that they are effectively watertight.

2.8.2 Openings other than hatchways, machinery space openings, manholes and flush scuttles in the working or superstructure deck should be protected by enclosed structures fitted with weathertight doors or their equivalent. Companionways should be situated as close as practicable to the centreline of the vessel.

2.9 Ventilators

2.9.1 In vessels of 45 m in length and over, the height above deck of ventilator coamings, other than machinery space ventilator coamings, should be at least 900 mm on the working deck and at least 760 mm on the superstructure deck. In vessels of less than 45 m in length, the height of these coamings should be 760 mm and 450 mm respectively. The height above deck of machinery space ventilator openings should be to the satisfaction of the Competent Authority.

2.9.2 Coamings of ventilators should be of equivalent strength to the adjacent structure and capable of being closed weathertight by closing appliances permanently attached to the ventilator or adjacent structure. Where the coaming of any ventilator exceeds 900 mm in height it should be specially supported.

2.9.3 Closing appliances in vessels of 45 m in length and over need not be fitted to ventilators the coamings of which extend to more than 4.5 m above the working deck or more than 2.3 m above the superstructure deck unless specifically required by the Competent Authority. In vessels of less than 45 m in length, closing appliances need not be fitted to ventilators the coamings of which extend to more than 3.4 m above the working deck or more than 1.7 m above the superstructure deck. If the Competent Authority is satisfied that it is unlikely that water will enter the vessel through machinery space ventilators, closing appliances to such ventilators may be omitted.

2.10 Air pipes

2.10.1 Where air pipes to tanks and void spaces below deck extend above the working or the superstructure decks, the exposed parts of the pipes should be of strength equivalent to the adjacent structures and fitted with appropriate protection. Openings of air pipes should be provided with means of closing permanently attached to the pipe or adjacent structure.

2.10.2 The height of air pipes above deck to the point where water may have access below should be at least 760 mm on the working deck and at least 450 mm on the superstructure deck. The Competent Authority may accept reduction of the height of an air pipe to avoid interference with the fishing operations.

2.11 Sounding devices

2.11.1 Sounding devices, to the satisfaction of the Competent Authority, should be fitted:

    .1   to the bilges of those compartments which are not readily accessible at all times during the voyage; and

    .2   to all tanks and cofferdams.

2.11.2 Where sounding pipes are fitted, their upper ends should be extended to a readily accessible position and, where practicable, above the working deck. Their openings should be provided with permanently attached means of closing. Sounding pipes which are not extended above the working deck should be fitted with automatic self-closing devices.

2.11.3 Sounding arrangements on service fuel oil tanks should be such that, in the event of the tanks being overfilled, spillage through the means of sounding cannot occur.

2.11.4 Fuel oil tank sounding pipe openings should not be located in crew accommodation, but may exceptionally be installed in passageways, in which case flush deck screwed caps should be fitted.

2.12 Sidescuttles and windows

2.12.1 Sidescuttles to spaces below the working deck and to spaces within the enclosed structures on that deck should be fitted with hinged deadlights capable of being closed watertight.

2.12.2 No sidescuttle should be fitted in such a position that its sill is less than 500 mm above the deepest operating waterline.

2.12.3 Sidescuttles fitted less than 1000 mm above the deepest operating waterline should be of the fixed type.

Construction, watertight integrity and equipment

2.12.4 Sidescuttles, together with their glasses and deadlights, should be of an approved construction. Those prone to be damaged by fishing gear should be suitably protected.

2.12.5 Toughened safety glass or its equivalent should be used for the wheelhouse windows.

2.12.6 The Competent Authority may accept sidescuttles and windows without deadlights in side and aft bulkheads of deckhouses located on or above the working deck if satisfied that the safety of the vessel will not be impaired.

2.13 Inlets and discharges

2.13.1 Discharges led through the shell either from spaces below the working deck or from within enclosed superstructures or deckhouses on the working deck fitted with doors complying with the requirements of 2.4 should be fitted with accessible means for preventing water from passing inboard. Normally each separate discharge should have an automatic non-return valve with a positive means of closing it from an accessible position. Such a valve is not required if the Competent Authority considers that the entry of water into the vessel through the opening is not likely to lead to dangerous flooding and that the thickness of the piping is sufficient. The means for operating the positive-action valve should be provided with an indicator showing whether the valve is open or closed.

2.13.2 In manned machinery spaces, main and auxiliary sea inlets and discharges essential for the operation of machinery may be controlled locally. The controls should be accessible and should be provided with indicators showing whether the valves are open or closed.

2.13.3 Fittings attached to the shell and the valves required by 2.13.1 should be of steel, bronze or other approved ductile material. All pipes between the shell and the valves should be of steel, except that in spaces other than machinery spaces of vessels constructed of material other than steel the Competent Authority may approve the use of other materials.

2.14 Freeing ports

2.14.1 Where bulwarks on weather parts of the working deck form wells, the minimum freeing port area (A) in m^2, on each side of the vessel for each well on the working deck should be determined in relation to the length (l) and height of bulwark in the well as follows:

    .1 $A = 0.07l$

        (l need not be taken as greater than 0.7L).

    .2.1 Where the bulwark is more than 1200 mm in average height, the required area should be increased by 0.004 m^2 per metre of length of well for each 100 mm difference in height.

.2.2 Where the bulwark is less than 900 mm in average height, the required area may be decreased by 0.004 m^2 per metre of length of well for each 100 mm difference in height.

2.14.2 The freeing port area calculated according to 2.14.1 should be increased where the Competent Authority considers that the vessel's sheer is not sufficient to ensure that the deck is rapidly and effectively freed of water.

2.14.3 Subject to the approval of the Competent Authority, the minimum freeing port area for each well on the superstructure deck should be not less than one-half the area (A) given in 2.14.1.

2.14.4 Freeing ports should be so arranged along the length of bulwarks as to ensure that the deck is freed of water most rapidly and effectively. Lower edges of freeing ports should be as near the deck as practicable.

2.14.5 Poundboards and means for stowage of the fishing gear should be arranged so that the effectiveness of freeing ports will not be impaired. Poundboards should be so constructed that they can be locked in position when in use and should not hamper the discharge of shipped water.

2.14.6 Freeing ports over 300 mm in depth should be fitted with bars spaced not more than 230 mm nor less than 150 mm apart or provided with other suitable protective arrangements. Freeing port covers, if fitted, should be of approved construction. If devices are considered necessary for locking freeing port covers during fishing operations, they should be to the satisfaction of the Competent Authority and easily operable from a readily accessible position.

2.14.7 In vessels intended to operate in areas subject to icing, covers and protective arrangements for freeing ports should be capable of being easily removed to restrict ice accretion. The size of openings and means provided for removal of these protective arrangements should be to the satisfaction of the Competent Authority.

2.15   Anchor and mooring equipment

Anchor equipment designed for quick and safe operation should be provided which should consist of anchoring equipment, anchor chains or wire ropes, stoppers and a windlass or other arrangements for dropping and hoisting the anchor and for holding the vessel at anchor in all foreseeable service conditions. Vessels should also be provided with adequate mooring equipment for safe mooring in all operating conditions. Anchor and mooring equipment should be to the satisfaction of the Competent Authority. Recommended practice for anchor and mooring equipment is given in annex II.

2.16 Working deck within an enclosed superstructure

2.16.1 Such decks should be fitted with an efficient drainage system having an appropriate drainage capacity to dispose of washing water and fish offal.

2.16.2 All openings necessary for fishing operations should be provided with means for quick and efficient closure by one person.

2.16.3 Where the catch is brought onto such decks for handling or processing, the catch should be placed in a pound. Such pounds should comply with 3.11. An efficient drainage system should be fitted. Adequate protection against inadvertent influx of water to the working deck should be provided.

2.16.4 At least two exits from such decks should be provided.

2.16.5 The clear headroom in the working space should, at all points, be to the satisfaction of the Competent Authority.

2.16.6 A fixed ventilation system providing sufficient changes of air per hour should be provided.

2.17 Tanks for fish in refrigerated (RSW) or chilled (CSW) seawater

2.17.1 If RSW or CSW tanks or similar tank systems are used, such tanks should be provided with a separate, permanently fitted arrangement for the filling and emptying of seawater.

2.17.2 If such tanks are to be used also for carrying dry cargo, the tanks should be arranged with a bilge system and provided with adequate means to avoid ingress of water from the bilge system into the tanks.

Chapter III

Stability and associated seaworthiness

3.1 General

3.1.1 Vessels should be so designed and constructed that the requirements of this chapter will be satisfied in the operating conditions referred to in 3.7. Calculations of the righting lever curves should be to the satisfaction of the Competent Authority.*

3.1.2 Wherever practicable, guidance should be provided for an approximate determination of the vessel's stability by means of the rolling period test including values of rolling coefficients particular to the vessel. A suggested form for such guidance is shown at the appendix to the Memorandum to Administrations in this respect reproduced at appendix 7 of the annex to part A of the Code.

3.2 Stability criteria

3.2.1 The following minimum stability criteria should be applied unless the Competent Authority is satisfied that operating experience justifies departure therefrom:

> .1 The area under the righting lever curve (GZ curve) should not be less than 0.055 m·rad up to 30° angle of heel and not less than 0.090 m·rad up to 40° or the angle of flooding θ_f if this angle is less than 40°. Additionally, the area under the GZ curve between the angles of heel of 30° and 40° or between 30° and θ_f, if this angle is less than 40° should not be less than 0.030 m·rad. θ_f is the angle of heel at which openings in the hull, superstructures or deckhouses which cannot be closed water-

---

* Refer to the Calculation of stability curves and to the Effect of free surfaces of liquids in tanks (contained in paragraphs 3.6 and 3.3, respectively, of the Code on Intact Stability for all Types of Ships Covered by IMO Instruments, adopted by the Organization by resolution A.749(18), as amended), and the Code of Practice concerning the Accuracy of Stability Information for Fishing Vessels, adopted by the Organization by resolution A.267(VIII).

Stability and associated seaworthiness

tight commence to immerse. In applying this criterion, small openings through which progressive flooding cannot take place need not be considered as open;

.2 the righting lever GZ should be at least 200 mm at an angle of heel equal to or greater than 30°;

.3 the maximum righting lever GZ_{max} should occur at an angle of heel preferably exceeding 30° but not less than 25°; and

.4 the initial metacentric height GM_0 should not be less than 350 mm for single-deck vessels. In vessels with complete superstructure or vessels of 70 m in length and over, the metacentric height may be reduced to the satisfaction of the Competent Authority but in no case should be less than 150 mm.

3.2.2 Where arrangements other than a bilge keel are provided to limit the angles of roll, the Competent Authority should be satisfied that the stability criteria given in 3.2.1 are maintained in all operating conditions.

3.2.3 Where ballast is provided to ensure compliance with 3.2.1, its nature and arrangement should be to the satisfaction of the Competent Authority.

3.2.4 It should be ensured that stability characteristics of the vessel will not produce acceleration forces which could be prejudicial to the safety of the vessel and crew.

3.2.5 For a vessel with L less than 30 m for which, by reason of insufficient stability data, 3.2.1 cannot be applied, the following formula for the minimum metacentric height GM_{min}, in metres, for all operating conditions should be used as the criterion:*

$$GM_{min} = 0.53 + 2B\left[0.075 - 0.37\left(\frac{f_{min}}{B}\right) + 0.82\left(\frac{f_{min}}{B}\right)^2 - 0.014\left(\frac{B}{D}\right) - 0.032\left(\frac{l_s}{L}\right)\right]$$

where:

L, B, D and f_{min}, in metres, are as defined in 1.2.20, 1.2.5, 1.2.11 and 1.2.16 respectively; and

l_s = Actual length, in metres, of an enclosed superstructure, extending from side to side of the vessel, as defined in 1.2.29.

The formula is applicable for vessels having:

.1 $\dfrac{f_{min}}{B}$ between 0.02 and 0.20;

.2 $\dfrac{l_s}{L}$ smaller than 0.60;

---

* Refer to the Recommendation for an interim simplified stability criterion for decked fishing vessels under 30 m in length, contained in paragraph 4.2.6 of the Code on Intact Stability for all Types of Ships covered by IMO Instruments, adopted by the Organization by resolution A.749(18), as amended.

Chapter III

    .3 $\dfrac{B}{D}$ between 1.75 and 2.15;

    .4   sheer fore and aft at least equal to or exceeding the standard sheer prescribed in regulation 38(8) of the International Convention on Load Lines, 1966;

    .5   height of superstructure included in the calculation not less than 1.8 m.

For vessels with parameters outside of the above limits, the formula should be applied with special care.

3.2.6 The above formula is not intended as a replacement for the basic criteria given in 3.2.1 and 3.5 but should be used only if circumstances are such that cross-curves of stability, KM curve and subsequent GZ curves are not and cannot be made available for judging a particular vessel's stability.

3.2.7 The calculated value of GM_{min} should be compared with actual GM values of the vessel in all loading conditions. If a rolling test (see appendix 7 of the annex to part A of the Code), an inclining experiment based on estimated displacement, or another approximate method of determining the actual GM is used, a safety margin should be added to the calculated GM_{min}.

3.3   Flooding of fish-holds

The angle of heel at which progressive flooding of fish-holds could occur through hatches which remain open during fishing operations and which cannot rapidly be closed should be at least 20° unless the stability criteria of 3.2.1 can be satisfied with the respective fish-holds partially or completely flooded.

3.4   Particular fishing methods

Vessels engaged in particular fishing methods where additional external forces are imposed on the vessel during fishing operations should meet the stability criteria of 3.2.1 increased, if necessary, to the satisfaction of the Competent Authority.

3.5   Severe wind and rolling

Vessels should be able to withstand, to the satisfaction of the Competent Authority, the effect of severe wind and rolling in associated sea conditions, taking account of the seasonal weather conditions, the sea states in which the vessel will operate, the type of vessel and its mode of operation.*

3.6 Water on deck

Vessels should be able to withstand, to the satisfaction of the competent authority, the effect of water on deck, taking account of the seasonal weather conditions, the sea states in which the vessel will operate, the type of vessel and its mode of operation.†

3.7 Operating conditions

3.7.1 The number and type of operating conditions to be considered should be to the satisfaction of the Competent Authority and should include the following:

.1 departure for the fishing grounds with full fuel, stores, ice, fishing gear, etc.;

.2 departure from the fishing grounds with full catch;

.3 arrival at home port with full catch and 10% stores, fuel, etc.; and

.4 arrival at home port with 10% stores, fuel, etc. and a minimum catch, which should normally be 20% of full catch but may be up to 40% provided the Competent Authority is satisfied that operating patterns justify such a value.

3.7.2 In addition to the specific operating conditions given in 3.7.1, the Competent Authority should also be satisfied that the minimum stability criteria given in 3.2 are met under all other actual operating conditions, including those which produce the lowest values of the stability parameters contained in these criteria. The Competent Authority should also be satisfied that those special conditions associated with a change in the vessel's mode or areas of operation which affect the stability considerations of this chapter are taken into account.

3.7.3 Concerning the conditions referred to in 3.7.1, the calculations should include the following:

.1 allowance for the weight of the wet fishing nets and tackle, etc. on deck;

---

* Refer to the Severe wind and rolling criterion (weather criterion) for fishing vessels, contained in paragraph 4.2.4 of the Code on Intact Stability for all Types of Ships Covered by IMO Instruments adopted by the Organization by resolution A.749(18), as amended.

† Refer to the Guidance on a method of calculation of the effect of water on deck, contained in recommendation 1 of attachment 3 to the Final Act of the 1993 Torremolinos Conference.

.2 allowance for ice accretion, if anticipated, in accordance with 3.8;

.3 homogeneous distribution of the catch, unless this is inconsistent with practice;

.4 catch on deck, if anticipated, in operating conditions referred to in 3.7.1.2, 3.7.1.3 and 3.7.2;

.5 water ballast, if carried either in tanks which are especially provided for this purpose or in other tanks also equipped for carrying water ballast; and

.6 allowance for the free surface effect of liquids and, if applicable, catch carried.

3.8 Ice accretion

3.8.1 For vessels operating in areas where ice accretion is likely to occur, the following icing allowance should be made in the stability calculations:*

.1 30 kg/m^2 on exposed weather decks and gangways;

.2 7.5 kg/m^2 for the projected lateral area of each side of the vessel above the water plane; and

.3 the projected lateral area of discontinuous surfaces of rail, spars (except masts) and rigging of vessels having no sails and the projected lateral area of other small objects should be computed by increasing the total projected area of continuous surfaces by 5% and the static moments of this area by 10%.

3.8.2 The height of the centre of gravity of ice accretion should be calculated according to the position of corresponding parts of the decks and gangways and other continuous surfaces on which ice can accumulate.

3.8.3 Vessels intended for operation in areas where ice accretion is known to occur should be:

.1 designed to minimize the accretion of ice; and

.2 equipped with such means for removing ice as the Competent Authority may require.†

---

* For sea areas where ice accretion may occur and modifications of the icing allowance are suggested, refer to the Guidance relating to ice accretion, contained in recommendation 2 of attachment 3 to the Final Act of the 1993 Torremolinos Conference. Refer also to the Icing consideration and the Recommendation for skippers of fishing vessels on ensuring a vessel's endurance in conditions of ice formation, contained in appendix 10 of the annex to part A of the Code.

† Refer to paragraph 2.4 of appendix 10 of the annex to part A of the Code on a typical list of equipment and hand tools required for combating ice formation.

3.9 Inclining test*

3.9.1 Every vessel should undergo an inclining test upon its completion and the actual displacement and position of the centre of gravity should be determined for the lightship condition.

3.9.2 Where alterations are made to a vessel affecting its lightship condition and the position of the centre of gravity, the vessel should, if the Competent Authority considers this necessary, be re-inclined and the stability information revised.

3.9.3 The Competent Authority may allow the inclining test of an individual vessel to be dispensed with, provided basic stability data are available from the inclining test of a sister ship, and it is shown to the satisfaction of the Competent Authority that reliable stability information for the exempted vessel can be obtained from such basic data.

3.10 Stability information

3.10.1 Suitable stability information should be supplied to enable the skipper to assess with ease and certainty the stability of the vessel under various operating conditions.† Such information should include specific instructions to the skipper warning him of those operating conditions which could adversely affect either the stability or the trim of the vessel. A copy of the stability information should be submitted to the Competent Authority for approval.‡

3.10.2 The approved stability information should be kept on board, readily accessible at all times and inspected at the periodical surveys of the vessel to ensure that it has been approved for the actual operating conditions.

3.10.3 Where alterations are made to a vessel affecting its stability, revised stability calculations should be prepared and submitted to the Competent Authority for approval. If the Competent Authority decides that the stability information must be revised, the new information should be supplied to the skipper and the superseded information removed.

---

* Refer to the Determination of lightship displacement and centres of gravity and the Detailed guidance for the conduct of an inclining test, contained in chapter 7 and annex 1 respectively of the Code on Intact Stability for all Types of Ships Covered by IMO Instruments, adopted by the Organization by resolution A.749(18), as amended.

† Refer to the Guidance on stability information contained in recommendation 3 of attachment 3 to the Final Act of the 1993 Torremolinos Conference. See also the General provisions against capsizing and information for the master contained in chapter 2 of the Code on Intact Stability for all Types of Ships Covered by IMO Instruments adopted by the Organization by resolution A.749(18), as amended.

‡ Refer to the Code of practice concerning the accuracy of stability information for fishing vessels, adopted by the Organization by resolution A.267(VIII).

3.10.4 Scales indicating the vessel's draught should be permanently marked on both sides of the stem and stern. These scales should be measured perpendicularly from a datum line which will lie along, or be a projection of, the lower extremity of the keel or other appendage. Numbers 0.1 m in the vertical plane should be marked on the scale, the lower edge of each number indicating the draught in metres. Between the numbers, lines should be marked, parallel to the datum, at intervals of 0.1 m. The skipper should be provided with information defining the position of the datum line and instructions regarding the use of observed draughts.

3.11 Portable fish-hold divisions

The catch should be properly secured against shifting which could cause dangerous trim or heel of the vessel. Recommended practice on portable fish-hold divisions is given in annex III to this part of the Code. The scantlings of portable fish-hold divisions, if fitted, should be to the satisfaction of the Competent Authority.

3.12 Bow height

The bow height should be sufficient, to the satisfaction of the Competent Authority, to prevent the excessive shipping of water and should be determined taking account of the seasonal weather conditions, the sea states in which the vessel will operate, the type of vessel and its mode of operation.

3.13 Maximum permissible operating draught

3.13.1 A maximum permissible operating draught should be approved by the Competent Authority and should be such that, in the associated operating condition, the stability criteria of this chapter and the provisions of chapters II and VI as appropriate are satisfied.

3.13.2 The maximum permissible operating draught should be marked on each side of the vessel. Working deck should be marked by working deck line. The location of the maximum permissible operating draught mark and the working deck line should be indicated on one of the safety certificates for the vessel.

3.14 Subdivision and damage stability

Vessels of 100 m in length and over, where the total number of persons carried is 100 or more, should be capable, to the satisfaction of the Competent Authority, of remaining afloat with positive stability after the flooding of any one compartment assumed damaged, having regard to the type of vessel, the intended service and area of operation.*

---

* Refer to the Guidance on subdivision and damage stability calculations, contained in recommendation 5 of attachment 3 to the Final Act of the 1993 Torremolinos Conference.

Chapter IV

Machinery and electrical installations and periodically unattended machinery spaces

Part A
General

4.1 Definitions

4.1.1 *Auxiliary means of activating the rudder* is the equipment which is provided for effecting movement of the rudder for the purpose of steering the vessel in the event of failure of the main steering gear.

4.1.2 *Dead vessel condition* is the condition under which the main propulsion plant boilers and auxiliaries are not in operation due to the absence of power.

4.1.3 *Fuel oil unit* is the equipment used for the preparation of fuel oil for delivery to an oil-fired boiler, or equipment used for the preparation of oil for delivery to an internal combustion engine, and includes any oil pressure pumps, filters and heaters dealing with oil at a pressure greater than 0.18 N/mm^2.

4.1.4 *Main steering gear* is the machinery, the steering gear power units, if any, and ancillary equipment and the means of applying torque to the rudder stock (e.g. tiller or quadrant) necessary for effecting movement of the rudder for the purpose of steering the vessel under normal service conditions.

4.1.5 *Main switchboard* is a switchboard directly supplied by the main source of electrical power and intended to distribute electrical energy.

4.1.6 *Maximum ahead service speed* is the greatest speed which the vessel is designed to maintain in service at sea at its maximum permissible operating draught.

4.1.7 *Maximum astern speed* is the speed which it is estimated the vessel can attain at the designed maximum astern power at its maximum permissible operating draught.

4.1.8 *Normal operational and habitable conditions* means conditions under which the vessel as a whole, its machinery services, means of main and auxiliary propulsion, steering gear and associated equipment, aids to safe navigation and to limit the risks of fire and flooding, internal and external means of communicating and signalling, means of escape and winches for rescue boats, are in proper working order and the minimum comfortable conditions of habitability are satisfactory.

4.1.9 *Periodically unattended machinery spaces* means those spaces containing main propulsion and associated machinery and all sources of main electrical supply which are not at all times manned under all operating conditions, including manoeuvring.

4.1.10 *Steering gear power unit* means, in the case of:

.1 electric steering gear, an electric motor and its associated electrical equipment;

.2 electro-hydraulic steering gear, an electric motor and its associated electrical equipment and connected pump; and

.3 other hydraulic steering gear, a driving engine and connected pump.

4.2 General

Machinery installations

4.2.1 Main propulsion, control, steam and hydraulic pipes, fuel oil, compressed air, electrical and refrigeration systems, auxiliary machinery, boilers and other pressure vessels, piping and pumping management, steering equipment and gears, shafts and couplings for power transmission should be designed, constructed, tested, installed and serviced to the satisfaction of the Competent Authority. This machinery and equipment, as well as lifting gear, winches, fish handling and fish processing equipment, should be protected so as to reduce to a minimum any danger to persons on board. Special attention should be paid to moving parts, hot surfaces and other dangers.

4.2.2 Machinery spaces should be so designed as to provide safe and free access to all machinery and its controls as well as to any other parts which may require servicing. Such spaces should be adequately ventilated.

4.2.3 Passages of sufficient width but not less than 600 mm should be provided between main engine and auxiliary machinery or main switchboard.

4.2.4 Means should be provided whereby the operational capability of the propulsion machinery can be sustained or restored even though one of the essential auxiliaries becomes inoperative. Special consideration should be given to the functioning of:

.1 the arrangements which supply fuel oil pressure for main propulsion machinery;

.2 the normal sources of lubricating oil pressure;

.3 the hydraulic, pneumatic and electrical means for the control of main propulsion machinery, including controllable-pitch propellers;

.4 the sources of water pressure for main propulsion cooling systems; and

.5 an air compressor and an air receiver for starting or control purposes,

provided that the Competent Authority may, having regard to overall safety considerations, accept a partial reduction in capability in lieu of full normal operation.

4.2.5 To the extent practicable, means should be provided whereby the machinery can be brought into operation from dead vessel condition without external aid.

4.2.6 Main propulsion machinery and all auxiliary machinery essential to the propulsion and the safety of the vessel should, as fitted, be capable of operating whether the vessel is upright or listed up to 15° either way under static conditions and up to 22.5° either way under dynamic conditions, i.e. when rolling either way and simultaneously pitching (inclined dynamically) up to 7.5° by bow or stern. The Competent Authority may permit deviation from these angles, taking into consideration the type, size and service conditions of the vessel.

4.2.7 Special consideration should be given to the design, construction and installation of propulsion machinery systems so that any mode of their vibrations should not cause undue stresses in such machinery systems in the normal operating ranges.

4.2.8 All controls for operating the machinery and equipment, measuring devices, pumping systems and arrangements, valves, cocks, air pipes, inlets, sounders, switches, etc. should be permanently marked with appropriate inscriptions clearly showing their purpose. Pipes should preferably be marked by appropriate colours to indicate their purpose. All handwheels should be marked with pointers showing the direction of turning, which generally should be clockwise for closure.

4.2.9 Steam fittings, steam pipes and exhaust pipes and other hot surfaces within reach of personnel should be properly insulated or otherwise protected to prevent accidents or burns. Hot surfaces which could cause ignition should be protected from all possible contacts with combustible liquid.

4.2.10 Plastic piping should not be used for any purpose in the machinery spaces where its destruction by fire would present a safety hazard.

4.2.11 Railings on gratings in the machinery spaces should consist of a handrail and guard rail where practicable; toe boards approximately 60 mm high should be affixed to the edge of all gratings where appropriate.

4.2.12 Openings to machinery space bilges should be properly guarded with handrails and toe boards or gratings.

4.2.13 Floor plates should be properly fitted and secured in place and should have a non-slip surface where practicable.

4.2.14 Machinery space ladders should be fitted with non-slip treads and well maintained. Adequate handrails should be provided.

4.2.15 Spare parts and stores should be provided to the satisfaction of the Competent Authority. Adequate facilities should be provided for the safe stowage of spare parts and stores.

4.2.16 Information on operation and maintenance of machinery and boilers, usage of fuels and lubricating oils should be provided.

4.2.17 Water-level indicators, pressure gauges and other measuring devices should be so installed and illuminated as to be readily visible.

Electrical installations

4.2.18 The design and construction of electrical installations should be such as to provide:

    .1   the services necessary to maintain the vessel in normal operational and habitable conditions without having recourse to an emergency source of power;

    .2   the services essential to safety when failure of the main source of electrical power occurs; and

    .3   protection of the crew and vessel from electrical hazards.

4.2.19 The Competent Authority should be satisfied that the provisions of 4.15 to 4.17 are uniformly implemented and applied.

Periodically unattended machinery spaces

4.2.20 In addition to 4.2 to 4.17 and chapter V, the provisions of 4.18 to 4.23 should apply to vessels with periodically unattended machinery spaces.

4.2.21 Measures should be taken to the satisfaction of the Competent Authority to ensure that all equipment is functioning in a reliable manner in all operating conditions, including manoeuvring, and that arrangements to the satisfaction of the Competent Authority are made for regular inspections and routine tests to ensure continuous reliable operation.

Chapter IV: part B

4.2.22 Vessels should be provided with documentary evidence, to the satisfaction of the Competent Authority, of their fitness to operate with periodically unattended machinery spaces.

Part B
Machinery installations
(See also section 4.2)

4.3 Machinery

4.3.1 Main and auxiliary machinery essential for the propulsion and safety of the vessel should be provided with effective means of control.

4.3.2 Where main or auxiliary machinery, including pressure vessels or any parts of such machinery, are subject to internal pressure and may be subject to dangerous overpressure, means should be provided, where applicable, which will protect against such excessive pressure.

4.3.3 All gearing and every shaft and coupling used for transmission of power to machinery essential for the propulsion and safety of the vessel or the safety of persons on board should be so designed and constructed that it will withstand the maximum working stresses to which it may be subjected in all service conditions. Due consideration should be given to the type of engines by which it is driven or of which it forms part.

4.3.4 Main propulsion machinery and, where applicable, auxiliary machinery should be provided with automatic shut-off arrangements in the case of failures, such as lubricating oil supply failure, which could lead rapidly to damage, complete breakdown or explosion. An advance alarm should also be provided so that warning is given before automatic shut-off but the Competent Authority may permit provisions for overriding automatic shut-off devices. The Competent Authority may also permit relaxation of the provisions of this paragraph, giving consideration to the type of vessel or its specific service.

4.3.5 Bars used on flywheels to turn machinery over by hand should be so constructed as to facilitate easy withdrawal from the flywheel's recess if the engine should recoil. Hand cranks for engines should be designed to be thrown out instantly when the engine starts.

4.4 Means of going astern

4.4.1 Vessels should have sufficient power for going astern to secure proper control of the vessel in all normal circumstances.

4.4.2 The ability of the machinery to reverse the direction of thrust of the propeller in sufficient time and so to bring the vessel to rest within a reasonable distance from maximum ahead service speed should be demonstrated at sea.

4.5 Steam boilers, feed systems and steam piping arrangements

4.5.1 Every steam boiler and every unfired steam generator should be provided with not less than two safety valves of adequate capacity. However, the Competent Authority may, having regard to the output or any other features of any steam boiler or unfired steam generator, permit only one safety valve to be fitted if satisfied that adequate protection against overpressure is thereby provided.

4.5.2 Every oil-fired steam boiler which is intended to operate without manual supervision should have safety arrangements which shut off the fuel supply and give an alarm in the case of low water level, air supply failure or flame failure.

4.5.3 The Competent Authority should give special consideration to steam boiler installations to ensure that feed systems, monitoring devices, and safety provisions are adequate in all respects to ensure the safety of boilers, steam pressure vessels and steam piping arrangements.

4.5.4 Auxiliary boilers operating on fuel oil which are located outside any enclosures on the platforms or 'tween-decks should be protected with oil-tight coamings of a height of approximately 200 mm.

4.5.5 Copper pipes used in steam supply and exhaust systems should be seamless.

4.5.6 Main and auxiliary steam stop valves should be arranged to seat against boiler pressure.

4.5.7 Steam supply and exhaust pipes should not be led through coal bunkers or dry cargo spaces unless approved by the Competent Authority, in which case they should be substantially encased for protection against mechanical injury. In vessels built of material other than steel, steam supply and exhaust piping should be insulated so that materials adjacent thereto are neither ignited nor rendered ineffective by heat.

4.5.8 Where more than one power boiler is fitted, the auxiliary steam piping should be so arranged that steam for whistle, steering gear, and electric lighting plant can be supplied from any power boiler.

4.5.9 Suitable drains should be provided at low points of piping systems.

4.5.10 Where positive shut-off valves are fitted in exhaust lines of machinery and the exhaust lines are not designed for the maximum inlet pressure, relief valves of sufficient capacity should be fitted between machinery exhaust and shut-off valves.

Chapter IV: part B

4.5.11 A sentinel relief valve or other warning device fitted on the engine or turbine exhaust may be permitted by the Competent Authority as a substitute for a relief valve, provided that a backpressure trip device is installed which will close the inlet valve when the exhaust side of the system is subjected to pressure exceeding the maximum allowable working pressure.

4.5.12 Shore steam connections should, where necessary, be fitted with reduction and relief valves set at a pressure not exceeding the design pressure of the piping.

4.5.13 Hot water heating systems should be designed as independent systems and approved by the Competent Authority.

4.6 Communication between the wheelhouse and machinery space

Two separate means of communication between the wheelhouse and the machinery space control platform should be provided; one of the means should be an engine-room telegraph. However, in vessels of less than 45 m in length, where the propulsion machinery is directly controlled from the wheelhouse, the Competent Authority may accept only one means of communication other than an engine-room telegraph. Due account should be taken of the noise level in the engine-room when selecting and locating these means of communication.

4.7 Wheelhouse control of propulsion machinery

4.7.1 Where remote control of propulsion machinery is provided from the wheelhouse, the following should apply:

.1 under all operating conditions, including manoeuvring, the speed, direction of thrust and, if applicable, the pitch of the propeller should be fully controllable from the wheelhouse;

.2 the remote control referred to in 4.7.1.1 should be performed by means of a control device to the satisfaction of the Competent Authority with, where necessary, means of preventing overload of the propulsion machinery;

.3 the main propulsion machinery should be provided with an emergency stopping device in the wheelhouse and independent from the wheelhouse control system referred to in 4.7.1.1;

.4 remote control of the propulsion machinery should be possible only from one station at a time; at any control station interlocked control units may be permitted. There should be at each station an indicator showing which station is in control of the propulsion machinery. The transfer of control between the wheelhouse and machinery spaces should be possible only in the machinery space or control room. The Competent

Machinery and electrical installations

Authority may permit the control station in the machinery space to be an emergency station only, provided that the monitoring and control in the wheelhouse is adequate;

.5 for vessels of 37 m in length and over, indicators should be fitted in the wheelhouse for:

.5.1 propeller speed and direction in the case of fixed propellers;

.5.2 propeller speed and pitch position in the case of controllable-pitch propellers; and

.5.3 advance alarm as required in 4.3.4;

.6 it should be possible to control the propulsion machinery locally even in the case of failure in any part of the remote control system;

.7 unless the Competent Authority considers it impracticable, the design of the remote control system should be such that, if it fails, an alarm will be given and the pre-set speed and direction of thrust will be maintained until local control is in operation; and

.8 special arrangements should be provided to ensure that automatic starting should not exhaust the starting possibilities. An alarm should be provided to indicate low starting possibilities. An alarm should be provided to indicate low starting air pressure and should be set at a level which will still permit main engine starting operations.

4.7.2 Where the main propulsion and associated machinery, including sources of main electrical supply, are provided with various degrees of automatic or remote control and are under continuous manned supervision from a control room, the control room should be so designed, equipped and installed that the machinery operation will be as safe and effective as if it were under direct supervision.

4.7.3 In general, automatic starting, operational and control systems should include means for manually overriding the automatic means, even in the case of failure of any part of the automatic and remote control system.

4.8 Air pressure system

4.8.1 Means should be provided to prevent excess pressure in any part of compressed air systems and wherever water-jackets or casings of air compressors and coolers might be subjected to dangerous excess pressure due to leakage into them from air pressure parts. Suitable pressure-relief arrangements should be provided.

4.8.2 The main starting air arrangements for main propulsion internal-combustion engines, if their cylinder diameters are more than 300 mm, should be adequately protected against the effects of backfiring and internal explosion in the starting air pipes.

4.8.3 All discharge pipes from starting air compressors should lead directly to the starting air receivers and all starting pipes from the air receivers to main or auxiliary engines should be entirely separate from the compressor discharge pipe system.

4.8.4 Provision should be made to reduce to a minimum the entry of oil into the air pressure systems and to drain these systems.

4.8.5 Air intakes for air compressors should be so located that the air is as pure and clean as possible and free from inflammable or toxic gases or fumes. Air filters should be fitted. Air discharge pipes of compressors should, where necessary, be insulated to protect personnel from burns.

4.9 Arrangements for fuel oil, lubricating oil and other flammable oils

4.9.1 Fuel oil which has a flashpoint of less than 60°C (closed-cup test) as determined by an approved flashpoint apparatus should not be used as fuel, except in emergency generators, in which case the flashpoint should be not less than 43°C. Provided that the Competent Authority may permit the general use of fuel oil having a flashpoint of not less than 43°C subject to such additional precautions as it may consider necessary and on condition that the temperature of the space in which such fuel is stored or used should not rise to within 10°C below the flashpoint of the fuel.

4.9.2 Safe and efficient means of ascertaining the amount of fuel oil contained in any oil tank should be provided. If sounding pipes are installed, their upper ends should terminate in safe positions and should be fitted with suitable means of closure. Gauges made of glass of substantial thickness and protected with a metal case may be used, provided that automatic closing valves are fitted. Other means of ascertaining the amount of fuel oil contained in any fuel oil tank may be permitted, providing their failure or overfilling of the tanks will not permit release of fuel.

4.9.3 Provision should be made to prevent overpressure in any oil tank or in any part of the fuel oil system including the filling pipes. Relief valves and air or overflow pipes should discharge to a position and in a manner which is safe.

4.9.4 Subject to the satisfaction of the Competent Authority, fuel oil pipes which, if damaged, would allow oil to escape from a storage, settling or daily service tank situated above the double bottom should be fitted with a cock or valve on the tank capable of being closed from a safe position outside the space concerned in the event of a fire arising in the space in which such tanks are situated.

Machinery and electrical installations

4.9.5 Pumps forming part of the fuel oil system should be separated from any other system and the connections of any such pumps should be provided with an efficient relief valve which should be in closed circuit. Where fuel oil tanks are alternatively used as liquid ballast tanks, proper means should be provided to isolate the fuel oil and ballast systems.

4.9.6 No oil tank should be situated where spillage or leakage therefrom can constitute a hazard by falling on heated surfaces. Precautions should be taken to prevent any oil that may escape under pressure from any pump, filter or heater from coming into contact with heated surfaces.

4.9.7 Fuel oil pipes and their valves and fittings should be steel or other equivalent material, provided that restricted use of flexible pipes may be permitted in positions where the Competent Authority is satisfied that they are necessary. Such flexible pipes and end attachments should be of adequate strength and should, to the satisfaction of the Competent Authority, be constructed of approved fire-resistant materials or have fire-resistant coatings. Where necessary, fuel oil and lubricating oil pipelines should be screened or otherwise suitably protected to avoid, as far as practicable, oil spray or oil leakage on heated surfaces or into machinery air intakes. The number of joints in piping systems should be kept to a minimum.

4.9.8 Fuel pipes of internal-combustion engines should be of steel of other equivalent material and preferably of a jacketed design. All fuel pipes should be adequately secured and protected.

4.9.9 As far as practicable, fuel oil tanks should be part of the vessel's structure and should be located outside machinery spaces of category A. Where fuel oil tanks, other than double-bottom tanks, are necessarily located adjacent to or within machinery spaces of category A, at least one of their vertical sides should be contiguous to the machinery space boundaries, and should preferably have a common boundary with the double-bottom tanks where fitted and the area of the tank boundary common with the machinery space should be kept to a minimum. For fuel oil tanks having capacity of not more than 15-hour continuous running of the vessel at maximum continuous rating of main propulsion machinery, the above requirements are relaxed. When such tanks are sited within the boundaries of machinery spaces of category A, if the vessel is not less than 37 m in length or the capacity of the tank is more than 1 m^3, they should not contain fuel oil having a flashpoint of less than 60°C (closed-cup test). In general, the use of freestanding fuel oil tanks should be avoided in fire hazard areas, and particularly in machinery spaces of category A. When freestanding fuel oil tanks are permitted, they should be placed in an oil-tight spill tray of ample size having a suitable drain pipe leading to a suitably sized spill oil tank.

4.9.10 The ventilation of machinery spaces should be sufficient under all normal conditions to prevent accumulation of oil vapour.

4.9.11 The arrangements for the storage, distribution and use of oil employed in pressure lubrication systems should be to the satisfaction of the Competent Authority. Such arrangements in machinery spaces of category A and, wherever practicable, in other machinery spaces should at least comply with the provisions of 4.9.1, 4.9.3, 4.9.6 and 4.9.7 and, insofar as the Competent Authority may consider necessary, with those of 4.9.2 and 4.9.4. This does not preclude the use of sight flow glasses in lubrication systems provided they are shown by test to have a suitable degree of fire resistance.

4.9.12 The arrangements for the storage, distribution and use of flammable oils employed under pressure in power transmission systems other than oils referred to in 4.9.11 in control and activating systems and heating systems should be to the satisfaction of the Competent Authority. In locations where means of ignition are present, such arrangements should at least comply with the provisions of 4.9.2 and 4.9.6 and with the provisions of 4.9.3 and 4.9.7 in respect of strength and construction.

4.9.13 Fuel oil, lubricating oil and other flammable oils should not be carried in forepeak tanks.

4.9.14 Oil storage tanks should not be situated above stairways and ladders, boilers, hot surfaces and electrical equipment. Oil storage tanks and piping should be arranged to minimize the possibility, in the event of overflow, leakage or rupture, that fuel will come into contact with hot surfaces or electrical components which may cause ignition of the fuel.

4.9.15 Vent pipes from fuel oil tanks should have a net cross-section not less than 1.25 times that of the filling pipes, and should be led from the top of the tank to the open air in a space where no danger will result from overflow or the discharge of oil vapours. The vent pipe outlets should be fitted with U bends (or other protective arrangements) and metal flame screens easily removable for cleaning. The open area of the screens should be not less than the cross-section area of the vent pipe.

4.9.16 Fuel oil tank sounding pipe openings should not be located in crew accommodation, but may exceptionally be installed in passageways, in which case flush deck screwed caps should be fitted.

4.9.17 Fuel oil filling stations should be outside the machinery spaces and so arranged that any overflow cannot come into contact with any hot surface where it might be ignited.

4.9.18 Removable oil burners of boilers should be so constructed as to be removable only after the burner's fuel oil valve has been closed. To ensure the correct sequence for turning on and off fuel oil burners of boilers, fuel oil valves and air dampers should be so arranged that fuel oil inlet valves can be opened only after air inlet dampers have been opened, and that in turning off oil burners, air inlet dampers can be closed only after fuel oil inlet valves have been secured. Fuel oil filters should be so placed as to minimize the danger of spraying oil on to hot surfaces and it should not be possible to remove the covers of any such filter until that filter has been properly isolated from the supply.

4.9.19 Overflow pipes from daily service tanks/settling tanks to double-bottom and/or bunker tanks should be fitted with a sight glass and an audible alarm.

4.10 Bilge pumping arrangements

4.10.1 An efficient bilge pumping plant should be provided which, under all practical conditions, should be capable of pumping from and draining any watertight compartment which is neither a permanent oil tank nor a permanent water tank, whether the vessel is upright or listed. Arrangements should be provided for easy flow of water to the suction pipes. Provided the Competent Authority is satisfied that the safety of the vessel is not impaired, the bilge pumping arrangements may be dispensed with in particular compartments.

4.10.2 At least two independently driven bilge pumps should be provided, one of which may be driven by the main engine. A ballast pump or other general service pump of sufficient capacity may be used as a power-driven bilge pump.

Power bilge pumps should be capable of giving a speed of water of at least 2 m/s through the main bilge pipe, which should have an internal diameter of at least:

$$d = 25 + 1.68\sqrt{L(B+D)}$$

where d is the internal diameter in millimetres, and L, B and D are in metres.

However, the actual internal diameter of the bilge main may be rounded off to the nearest standard size acceptable to the Competent Authority. In no case should the capacity of the bilge pump(s) be less than the capacity of the installed fire pump(s).

4.10.3 No bilge suction should have an inside diameter of less than 50 mm. The arrangement and sizing of the bilge system should be such that the full rated capacity of the pump specified above can be applied to each of the watertight compartments located between the collision and afterpeak bulkheads.

4.10.4 A bilge ejector in combination with an independently driven high-pressure seawater pump may be installed as a substitute for one independently driven bilge pump required by 4.10.2, provided this arrangement is to the satisfaction of the Competent Authority.

4.10.5 In vessels where fish handling or processing may cause quantities of water to accumulate in enclosed spaces, adequate drainage should be provided.

4.10.6 Bilge pipes should not be led through fuel oil, ballast or double-bottom tanks, unless these pipes are of heavy gauge steel construction.

4.10.7 Bilge and ballast pumping systems should be arranged so as to prevent water passing from the sea or from water ballast spaces into holds or into machinery spaces or from one watertight compartment to another. The bilge connection to any pump which draws from the sea or from water ballast spaces should be fitted with either a non-return valve or a cock which cannot be opened simultaneously either to the bilges and to the sea or to the bilges and water ballast spaces. Valves in bilge distribution boxes should be of a non-return type.

4.10.8 Any bilge pipe piercing a collision bulkhead should be fitted with a positive means of closing at the bulkhead with remote control from the working deck with an indicator showing the position of the valve, provided that, if the valve is fitted on the after side of the bulkhead and is readily accessible under all service conditions, the remote control may be dispensed with.

4.10.9 Valves and cocks not forming part of a piping system should not be permitted in watertight bulkheads.

4.10.10 Bilge suctions should be fitted with suitable strainers having an open area not less than three times the area of the bilge pipe.

4.10.11 One of the bilge pumps should have direct bilge suction from the compartment where the pump is situated.

4.10.12 In vessels of 45 m in length and over, the largest available power water pump in the engine-room suitable for use as a bilge pump should be fitted with an emergency bilge suction.

4.10.13 If fuel tanks are used to carry water ballast for ensuring stability or trim of the vessel, reliable devices should be provided for cutting off the ballast system from the tanks containing fuel as well as for cutting off the fuel system from fuel tanks containing water.

4.11 Protection against noise

Measures should be taken to reduce the effects of noise upon personnel in machinery spaces to levels satisfactory to the Competent Authority.

4.12 Steering gear

4.12.1 Vessels should be provided with a main steering gear and an auxiliary means of actuating the rudder to the satisfaction of the Competent Authority. The main steering gear and the auxiliary means of actuating the rudder should be arranged so that, so far as is reasonable and practicable, a single failure in one of them will not render the other one inoperative.

4.12.2 Where the main steering gear comprises two or more identical power units, an auxiliary steering gear need not be fitted if the main steering gear is capable of operating the rudder as required by 4.12.10 when any one of the units is out of operation. Each of the power units should be operated from a separate circuit.

4.12.3 The position of the rudder, if power-operated, should be indicated in the wheelhouse. The rudder angle indication for power-operated steering gear should be independent of the steering gear control system.

4.12.4 In the event of failure of the power-operated steering gear units, a visible and audible alarm should be given in the wheelhouse.

4.12.5 Indicators for running indication of the motors of electric and electrohydraulic steering gear should be installed in the wheelhouse. Short-circuit protection, an overload alarm and a no-voltage alarm should be provided for these circuits and motors. Protection against excess current, if provided, should be for not less than twice the full load current of the motor or circuit so protected, and should be arranged to permit the passage of the appropriate starting currents.

4.12.6 The main steering gear should be of adequate strength and sufficient to steer the vessel at maximum service speed. The main steering gear and rudder stock should be so designed that they will not be damaged at maximum speed astern or by manoeuvring during fishing operations.

4.12.7 The main steering gear should, with the vessel at its maximum permissible operating draught, be capable of putting the rudder over from 35° on one side to 35° on the other side with the vessel running ahead at maximum service speed. The rudder should be capable of being put over from 35° on either side to 30° on the other side in not more than 28 s, under the same conditions. The main steering gear should be operated by power, where necessary, to fulfil these requirements.

4.12.8 The main steering gear power unit should be arranged to start either by manual means in the wheelhouse or automatically when power is restored after a power failure.

4.12.9 The auxiliary means for actuating the rudder should be of adequate strength and sufficient to steer the vessel at navigable speed and capable of being brought speedily into action in an emergency.

4.12.10 The auxiliary means for actuating the rudder should be capable of putting the rudder over from 15° on one side to 15° on the other side in not more than 60 s with the vessel running at one half of its maximum service speed ahead or 7 knots, whichever is the greater. The auxiliary means for actuating the rudder should be operated by power where necessary to fulfil these requirements.

4.12.11 Electric or electro-hydraulic steering gear in vessels of 75 m in length and over should be served by at least two circuits fed from the main switchboard and these circuits should be as widely separated as possible.

Chapter IV: part B

4.13 Engineers' alarm

In vessels of 75 m in length and over, an engineers' alarm should be provided to be operated from the engine control room or at the manoeuvring platform as appropriate, and should be clearly audible in the engineers' accommodation.

4.14 Refrigeration systems for the preservation of the catch

4.14.1 Refrigeration systems should be so designed, constructed, tested and installed as to take account of the safety of the system and also the emission of refrigerants held in quantities or concentrations which are hazardous to human health or to the environment, and should be to the satisfaction of the Competent Authority.

4.14.2 Refrigerants to be used in refrigeration systems should be to the satisfaction of the Competent Authority. However, methyl chloride or CFCs whose ozone-depleting potential is higher than 5% of CFC-11 should not be used as refrigerants.

4.14.3 If ammonia is to be used as the refrigerant gas, the refrigerating plant should be at least arranged so as to take account of the recommended practice set out in annex IV to this part of the Code.

4.14.4 Refrigerating installations should be adequately protected against vibration, shock, expansion, shrinkage, etc. and should be provided with an automatic safety control device to prevent a dangerous rise in temperature and pressure.

4.14.5 Refrigeration systems in which toxic or flammable refrigerants are used should be provided with drainage devices leading to a place where the refrigerant presents no danger to the vessels or to persons on board.

4.14.6 Any space containing refrigerating machinery, including condensers and gas tanks utilizing toxic refrigerants, should be separated from any adjacent space by gastight bulkheads. Any space containing the refrigerating machinery, including condensers and gas tanks, should be fitted with a leak detection system having an indicator outside the space adjacent to the entrance and should be provided with an independent ventilation system.

4.14.7 Spaces containing condensers, gas tanks and refrigeration machinery utilizing toxic refrigerants, such as ammonia, should be provided with a water spray system.

4.14.8 When it is not practicable to contain refrigeration machinery in a separate place due to the size of the vessel, the refrigeration system may be installed in the machinery space provided that the quantity of refrigerant used will not cause danger to persons in the machinery space, should all the gas escape, and provided that an alarm is fitted to give warning of a dangerous concentration of gas should any leakage occur in the compartment.

4.14.9 In refrigerating machinery spaces and refrigerating rooms, alarms should be connected to the wheelhouse or control stations or escape exits to prevent persons being trapped. At least one exit from each such space should be capable of being opened from the inside. Where practicable, exits from the spaces containing refrigerating machinery using toxic or flammable gas should not lead directly into any accommodation spaces.

4.14.10 Where any refrigerant harmful to persons is used in a refrigeration system, at least two sets of breathing apparatus should be provided, one of which should be placed in a position not likely to become inaccessible in the event of leakage of refrigerant. Breathing apparatus provided as part of the vessel's fire-fighting equipment may be considered as meeting all or part of this provision provided its location meets both purposes. Where self-contained breathing apparatus is used, spare cylinders should be provided.

4.14.11 Adequate guidance for the safe operation and emergency procedures for the refrigeration system should be provided by suitable notices displayed on board the vessel.

Part C
Electrical installations
(See also section 4.2)

4.15 Main source of electrical power

Where electrical power constitutes the only means of maintaining auxiliary services essential for the propulsion and the safety of the vessel, a main source of electrical power should be provided which should include at least two generating sets, one of which may be driven by the main engine. The Competent Authority may accept other arrangements having equivalent electrical capability.

4.16 Emergency source of electrical power

4.16.1 A self-contained emergency source of electrical power should be provided and located outside the machinery spaces above the main deck. It should be so arranged as to ensure that it would function in the event of fire or other causes of failure of the main electrical installations.

4.16.2 The emergency source of electrical power, which may be either a generator or an accumulator battery, should be capable, having regard to starting current and the transitory nature of certain loads, of serving simultaneously, for a period of at least 3 h:

.1 a VHF radio installation or an MF radio installation or a ship-earth station or an MF/HF radio installation, depending on the sea area for which the vessel is to be equipped;

.2 internal communication equipment, fire detecting systems and signals which may be required in an emergency;

.3 the navigational lights, if solely electrical, and the emergency lights:

.3.1 at launching stations and over the side of the vessel;

.3.2 in all alleyways, stairways and exits;

.3.3 in spaces containing machinery or the emergency source of power;

.3.4 in control stations; and

.3.5 in fish-handling and fish-processing spaces;

.4 the operation of the emergency fire pump, if any.

4.16.3 The arrangements for the emergency source of electrical power should comply with the following:

.1 Where the emergency source of electrical power is a generator, it should be provided with an independent fuel supply and with efficient starting arrangements. Unless a second independent means of starting the emergency generator is provided, the single source of stored energy should be protected to preclude its complete depletion by the automatic starting system.

.2 Where the emergency source of electrical power is an accumulator battery, it should be capable of carrying the emergency load without recharging whilst maintaining the voltage of the battery throughout the discharge period within plus or minus 12% of its nominal voltage. In the event of failure of the main power supply, this accumulator battery should be automatically connected to the emergency switchboard and should immediately supply at least those services specified in 4.16.2.2 and 4.16.2.3. The emergency switchboard should be provided with an auxiliary switch allowing the battery to be connected manually in case of failure of the automatic connection system.

4.16.4 The emergency switchboard should be installed as near as is practicable to the emergency source of power and should be located in accordance with 4.16.1. Where the emergency source of power is a generator, the emergency switchboard should be located in the same place, unless the operation of the emergency switchboard would thereby be impaired.

4.16.5 Any accumulator battery should be installed in a well-ventilated space, but not in the space containing the emergency switchboard. An indicator should be mounted in a suitable place on the main switchboard or in the machinery control room to indicate when the battery constituting the emergency source of power is being discharged. The emergency switchboard should be supplied in normal operation from the main switchboard by an inter-connector feeder protected at the main switchboard against overload and short circuit. The arrangement at the emergency switchboard should be such that, in the event of a failure of the main power supply, an automatic connection of emergency supply would be provided. When the system is arranged for feedback operation, the inter-connector feeder should also be protected at the emergency switchboard against short circuit.

4.16.6 An emergency generator and its prime mover and any accumulator battery should be so arranged as to ensure that they will function at full rated power when the vessel is upright and when rolling up to an angle of 22.5° either way and simultaneously pitching 10° by bow or stern, or is in any combination of angles within those limits.

4.16.7 Battery level indicators should be mounted in a highly visible position on the main switchboard or in the machinery control room to facilitate monitoring of the condition of batteries constituting the emergency source of supply as well as any batteries required for the starting of an independent, power-driven emergency generator.

4.16.8 The emergency source of electrical power and automatic starting equipment should be so constructed and arranged as to enable adequate testing to be carried out by the crew while the vessel is in operating condition.

4.17 Precautions against shock, fire and other hazards of electrical origin

4.17.1 Electric circuits should be clearly identified on switchboards.

4.17.2 Electrical equipment exposed to the weather should be protected from dampness and corrosion as well as mechanical damage.

4.17.3 Piping conveying steam or liquid should not be fitted above or in the vicinity of switchboards or other electrical equipment. Where such arrangements are unavoidable, provision should be made to prevent leakage damaging the equipment.

4.17.4 Exposed permanently fixed metal parts of electrical machines or equipment which are not intended to be "live", but which are liable under fault conditions to become "live" should be earthed (grounded) unless:

    .1 they are supplied at a voltage not exceeding 55 V direct current or 55 V root mean square, between conductors; autotransformers should not be used for the purpose of achieving this voltage; or

Chapter IV: part C

    .2   they are supplied at a voltage not exceeding 250 V by safety isolating transformers supplying one consuming device only; or

    .3   they are constructed taking into account the principle of double insulation.

4.17.5 Portable electrical equipment should operate at a safe voltage; exposed metal parts of such equipment which are not intended to have a voltage, but which may have such under fault conditions, should be earthed. The Competent Authority may require additional precautions for portable electric lamps, tools or similar apparatus for use in confined or exceptionally damp spaces, where particular risks due to conductivity may exist.

4.17.6 Main and emergency switchboards should be so arranged as to give easy access as may be needed to apparatus and equipment, without danger to attendants. The sides and backs and, where necessary, the fronts of switchboards, should be suitably guarded. Exposed "live" parts having voltages to earth exceeding a voltage to be specified by the Competent Authority should not be installed on the front of such switchboards. There should be non-conducting mats or gratings at the front and rear, where necessary.

4.17.7 In vessels of 75 m in length and over, the hull-return system of distribution should not be used for power, heating or lighting in general. However, the requirement does not preclude, under conditions approved by the Competent Authority, the use of:

    .1   impressed current cathodic protective system;

    .2   limited locally earthed systems; or

    .3   insulation level monitoring devices, provided the circulation current does not exceed 30 mA under the most unfavourable conditions.

Where the hull-return system is used, all final sub-circuits (all circuits fitted after the last protective device) should be two wires, and special precautions should be taken to the satisfaction of the Competent Authority.

4.17.8 Except as permitted by the Competent Authority in exceptional circumstances, all metal sheaths and armour of cables should be electrically continuous and should be earthed.

4.17.9 Where the cables are neither sheathed nor armoured and there might be a risk of fire in the event of an electrical fault, special precautions should be taken to the satisfaction of the Competent Authority.

4.17.10 All electrical cables should be at least of a flame-retardant type and should be so installed as not to impair their original flame-retarding properties. The Competent Authority may permit the use of special types of cables, when necessary for particular applications, such as radio frequency cables, which do not apply with the foregoing.

4.17.11 Lighting fittings should be arranged to prevent temperature rises which could damage the wiring and to prevent surrounding material from becoming excessively hot.

4.17.12 Wiring should be supported in such a manner as to avoid chafing or other damage and should not be located close to hot surfaces such as engine exhausts.

4.17.13 Each separate circuit should be protected against short circuit and also against overload to the satisfaction of the Competent Authority, except in accordance with 4.12 or where the Competent Authority may exceptionally otherwise permit.

4.17.14 The rating or appropriate setting of the overload protective device for each circuit should be permanently indicated at the location of the protective device.

4.17.15 The housing of an accumulator battery should be constructed and ventilated to the satisfaction of the Competent Authority.

4.17.16 Electrical and other equipment which may constitute a source of ignition of flammable vapours should not be permitted in these compartments except as permitted in 4.17.18.

4.17.17 An accumulator battery should not be located in accommodation spaces unless installed in a hermetically sealed container.

4.17.18 In spaces, where flammable mixtures are liable to collect, and in any compartment assigned principally to the containment of an accumulator battery, no electrical equipment should be installed unless the Competent Authority is satisfied that it is:

.1 essential for operational purposes;

.2 of a type which will not ignite the mixture concerned;

.3 appropriate to the space concerned; and

.4 appropriately certified for safe usage in the dusts, vapours or gases likely to be encountered.

4.17.19 Where a potential explosion risk exists in or near any space, all electrical equipment and fittings installed in those spaces should be either explosion-proof or intrinsically safe to the satisfaction of the Competent Authority.

4.17.20 Cable systems and electrical equipment should be so installed as to avoid or reduce interference with radio reception.

Part D
Periodically unattended machinery spaces

4.18 Fire safety

Fire prevention

4.18.1 Special consideration should be given to high-pressure fuel oil pipes. Where practicable, leakages from such piping systems should be collected in a suitable drain tank which should be provided with a high level alarm.

4.18.2 Where daily service fuel oil tanks are filled automatically or by remote control, means should be provided to prevent overflow spillages. Similar consideration should be given to other equipment which treats flammable liquids automatically (e.g. fuel oil purifier) which, whenever practicable, should be installed in a special space reserved for purifiers and their heaters.

4.18.3 Where fuel oil daily service tanks or settling tanks are fitted with heating arrangements, a high temperature alarm should be provided if the flashpoint of the fuel oil can be exceeded.

Fire detection

4.18.4 An approved fire detection system based on a self-monitoring principle and including facilities for periodical testing should be installed in machinery spaces.

4.18.5 The detection system should initiate both audible and visual alarm in the wheelhouse and in sufficient appropriate spaces to be heard and observed by persons on board, when the vessel is in harbour.

4.18.6 The fire detection system should be fed automatically from an emergency source of power if the main source of power fails.

4.18.7 Internal-combustion engines of 2,500 kW and over should be provided with crankcase oil mist detectors or engine-bearing temperature detectors or equivalent devices.

Fire fighting

4.18.8 A fixed fire-extinguishing system should be provided to the satisfaction of the Competent Authority, which should be in compliance with the provisions of 5.22, 5.40 and 5.57.

4.18.9 In vessels of 75 m in length and over, provisions should be made for immediate water delivery from the fire main system either by:

    .1   remote starting arrangements of one of the main fire pumps in the wheelhouse and the fire control station, if any; or

.2 permanent pressurization of the fire main system, due regard being paid to the possibility of freezing.*

4.18.10 The Competent Authority should be satisfied with the maintenance of the fire integrity of the machinery spaces, the location and centralization of the fire-extinguishing system controls, the shut-down arrangements, (e.g. ventilation, fuel pumps, etc.) and may require fire-extinguishing appliances and other fire-fighting equipment and breathing apparatus in addition to the relevant provisions of chapter V.

4.19 Protection against flooding

4.19.1 Bilges in machinery spaces should be provided with a high level alarm in such a way that the accumulation of liquids is detected at normal angles of trim and heel. The detection system should initiate an audible and visual alarm in the places where continuous watch is maintained.

4.19.2 The controls of any valve serving a sea inlet or discharge below the waterline or a bilge injection system should be so sited as to allow adequate time for operation in case of influx of water to the space.

4.20 Communications

In vessels of 75 m in length and over, one of the two separate means of communication referred to in 4.6 should be a reliable vocal communication. An additional reliable means of vocal communication should be provided between the wheelhouse and the engineers' accommodation.

4.21 Alarm system

4.21.1 An alarm system should be provided which should indicate any fault requiring attention.

4.21.2 The alarm system should be capable of sounding an audible alarm in the machinery space and should indicate visually each separate alarm function at a suitable position. However, the Competent Authority may permit the system to be capable of sounding and indicating visually each separate alarm function in the wheelhouse only.

4.21.3 Audible and visual alarms should be activated in the wheelhouse for any situation requiring action by the responsible person on watch or which should be brought to his attention.

4.21.4 The alarm system should, as far as is practicable, be designed on the fail-safe principle.

---

* Refer to the Guidance for precautions against freezing of fire mains, contained in recommendation 6 of attachment 3 to the Final Act of the 1993 Torremolinos Conference.

Chapter IV: part D

4.21.5 The alarm system should be:

    .1 continuously powered with automatic change-over to a stand-by power supply in case of loss of normal power supply; and

    .2 activated by failure of the normal power supply.

4.21.6 The alarm system should be able to indicate at the same time more than one fault and the acceptance of any alarm should not inhibit another alarm.

4.21.7 Acceptance at the position referred to in 4.21.2 of any alarm condition should be indicated at the positions where it was shown. Alarms should be maintained until they are accepted and the visual indications should remain until the fault has been corrected. All alarms should automatically reset when the fault has been rectified.

4.22 Special requirements for machinery, boiler and electrical installations

4.22.1 In vessels of 75 m in length and over, the main source of electrical power should comply with the applicable provisions of the Protocol.

4.22.2 Where required to be duplicated, other auxiliary machinery essential to propulsion should be fitted with automatic change-over devices allowing transfer to a standby machine. An alarm should be given on automatic change-over.

4.22.3 Automatic control and alarm systems should be provided as follows:

    .1 the control system should be such that, through the necessary automatic arrangements, the services needed for the operation of the main propulsion machinery and its auxiliaries are ensured;

    .2 means should be provided to keep the starting air pressure at the required level where internal-combustion engines are used for main propulsion;

    .3 an alarm system complying with 4.21 should be provided for all important pressures, temperatures, fluid levels, etc.; and

    .4 where appropriate, an adequate central position should be arranged with the necessary alarm panels and instrumentation indicating any alarmed fault.

4.23 Safety system

A safety system should be provided so that serious malfunction in machinery or boiler operations which presents an immediate danger should initiate the automatic shut-down of that part of the plant and an alarm should be given. Shut-down of the propulsion system should not be automatically activated except in cases which could lead to serious damage, complete breakdown, or explosion. Where arrangements for overriding the shutdown of the main propelling machinery are fitted, these should be such as to preclude inadvertent activation. Visual means should be provided to show whether or not it has been activated.

Chapter V

Fire protection, fire detection, fire extinction and fire fighting

Part A
General fire protection provisions

5.1 General

One of the following methods of protection should be adopted in accommodation and service spaces:

.1 Method IF – The construction of all internal divisional bulkheads of non-combustible "B" or "C" class divisions generally without the installation of a detection or sprinkler system in the accommodation and service spaces; or

.2 Method IIF – The fitting of an automatic sprinkler and fire alarm system for the detection and extinction of fire in all spaces in which fire might be expected to originate, generally with no restriction on the type of internal divisional bulkheads; or

.3 Method IIIF – The fitting of an automatic fire alarm and detection system in all spaces in which a fire might be expected to originate, generally with no restriction on the type of internal divisional bulkheads, except that in no case should the area of any accommodation space or spaces bounded by an "A" or "B" class division exceed 50 m^2. However, the Competent Authority could increase this area for public spaces The recommendations for the use of non-combustible materials in construction and insulation for the boundary bulkheads of machinery spaces, control stations, etc. and the protection of stairway enclosures and corridors should be common to all three methods.

5.2 Definitions

5.2.1 *"A" class divisions* are those divisions formed by bulkheads and decks, which comply with the following:

.1 they should be constructed of steel or other equivalent material;

Fire protection, detection, extinction and fighting

.2 they should be suitably stiffened;

.3 they should be so constructed as to be capable of preventing the passage of smoke and flame to the end of the one-hour standard fire test; and

.4 they should be insulated with approved non-combustible materials such that the average temperature of the unexposed side will not rise more than 140°C above the original temperature, nor will the temperature, at any one point, including any joint, rise more than 180°C above the original temperature, within the time listed below:

class "A-60"	60 min
class "A-30"	30 min
class "A-15"	15 min
class "A-0"	0 min

The Competent Authority may require a test of a prototype bulkhead or deck to ensure that it meets the above requirements for integrity and temperature rise in accordance with the Fire Test Procedures Code.

5.2.2 *Accommodation spaces* are those spaces used for public spaces, corridors, lavatories, cabins, offices, hospitals, cinemas, games and hobbies rooms and pantries containing no cooking appliances and similar spaces.

5.2.3 *"B" class divisions* are those divisions formed by bulkheads, decks, ceilings, or linings, which comply with the following:

.1 they should be so constructed as to be capable of preventing the passage of flame to the end of the first one-half hour of the standard fire test;

.2 they should have an insulation value such that the average temperature of the unexposed side will not rise more than 140°C above the original temperature, nor will the temperature at any one point, including any joint, rise more than 225°C above the original temperature, within the time listed below:

class "B-15"	15 min
class "B-0"	0 min; and

.3 they should be constructed of approved non-combustible materials and all materials entering into the construction and erection of "B" class divisions should be non-combustible with the exception that combustible veneers could be permitted provided they meet the relevant recommendations of this chapter.

The Competent Authority may require a test of a prototype division to ensure that it meets the above requirements for integrity and temperature rise in accordance with the Fire Test Procedures Code.

Chapter V: part A

5.2.4 *"C" class divisions* are those divisions constructed of approved non-combustible materials. They do not need to meet requirements relative to the passage of smoke and flame nor the limiting of temperature rise. Combustible veneers could be used provided they meet other requirements of this chapter.

5.2.5 *Continuous "B" class ceilings or linings* are those "B" class ceilings or linings which terminate only at an "A" or "B" class division.

5.2.6 *Control stations* are those spaces in which the vessel's radio or main navigation equipment or the emergency source of power is located, or where the fire recording or fire control equipment is centralized.

5.2.7 *"F" class divisions* are those divisions formed by bulkheads, decks, ceilings or linings which comply with the following:

.1 they should be so constructed as to be capable of preventing the passage of flame to the end of the first one-half hour of the standard fire test; and

.2 they should have an insulation value such that the average temperature of the unexposed side will not rise more than 140°C above the original temperature, nor will the temperature at any one point, including any joint, rise more than 225°C above the original temperature, up to the end of the first one-half hour of the standard fire test.

The Competent Authority may require a test of a prototype division to ensure that it meets the above requirements for integrity and temperature rise in accordance with the Fire Test Procedures Code.

5.2.8 *Fire Safety Systems Code* means the International Code for Fire Safety Systems, adopted by the Maritime Safety Committee of IMO by resolution MSC.98(73), as may be amended by IMO.

5.2.9 *Fire Test Procedures Code* means the International Code for Application of Fire Test Procedures, adopted by the Maritime Safety Committee of IMO by resolution MSC.61(67), as may be amended by IMO.

5.2.10 *Low flame-spread* means that the surface thus described will adequately restrict the spread of flame, to the satisfaction of the Competent Authority by an established test procedure.*

5.2.11 *Machinery spaces* are those machinery spaces of category A and all other spaces containing propulsion machinery, boilers, fuel oil units, steam and internal-combustion engines, generators, steering gear, major electrical machinery, oil filling stations, refrigerating, stabilizing, ventilating and air conditioning machinery and similar spaces, and trunks to such spaces.

---

* Refer to the Recommendation on improved fire test procedures for flammability of bulkheads, ceilings and deck finish materials, adopted by the Organization by resolution A.653(16).

5.2.12 *Machinery spaces of category A* are those spaces and trunks to such spaces, which contain internal-combustion type machinery, used either:

.1 for main propulsion; or

.2 for other purposes where such machinery has, in the aggregate, a total power output of not less than 750 kW,

or which contain any oil-fired boiler or fuel oil unit.

5.2.13 *Non-combustible material* means a material which neither burns nor gives off flammable vapours in sufficient quantity for self-ignition when heated to approximately 750°C, this being determined in accordance with the Fire Test Procedures Code. Any other material is a combustible material.

5.2.14 *Public spaces* are those portions of the accommodation spaces which are used for halls, dining rooms, lounges, and similar permanently enclosed spaces.

5.2.15 *Service spaces* are those spaces used for galleys, pantries containing cooking appliances, lockers and store-rooms, workshops other than those forming part of the machinery spaces and similar spaces and trunks to such spaces.

5.2.16 *A standard fire test* is one in which specimens of the relevant bulkheads or decks are exposed in a test furnace to temperatures corresponding approximately to the standard time–temperature curve specified in the Fire Test Procedures Code.

5.2.17 *Steel or other equivalent material* means steel or any material which, by itself or due to insulation provided, has structural and integrity properties equivalent to steel at the end of the applicable fire exposure to the standard fire test (e.g., aluminium alloy with appropriate insulation).

Part B
Fire safety measures in vessels of a length of 60 m and over

5.3 Structure

5.3.1 The hull, superstructure, structural bulkheads, decks and deckhouses should be constructed of steel or other equivalent material except as otherwise specified in 5.3.4.

5.3.2 The insulation of aluminium alloy components of "A" or "B" class divisions, except structures which, in the opinion of the Competent Authority, are non-load-bearing, should be such that the temperature of the structural core does not rise more than 200°C above the ambient temperature at any time during the applicable fire exposure to the standard fire test.

Chapter V: part B

5.3.3 Special attention should be given to the insulation of aluminium alloy components of columns, stanchions and other structural members required to support survival craft stowage, launching and embarkation areas, and "A" and "B" class divisions, to ensure:

.1 that for such members supporting survival craft areas and "A" class divisions, the temperature rise limitation specified in 5.3.2 should apply at the end of one hour; and

.2 that for such members required to support "B" class divisions, the temperature rise limitation specified in 5.3.2 should apply at the end of one half-hour.

5.3.4 Crowns and casings of machinery spaces of category A should be of steel construction adequately insulated and any openings therein should be suitably arranged and protected to prevent the spread of fire.

5.4 Bulkheads within the accommodation and service spaces

5.4.1 Within the accommodation and service spaces, all bulkheads required to be "B" class divisions should extend from deck to deck and to the shell or other boundaries. Unless continuous "B" class ceilings or linings, or both, are fitted on both sides of the bulkheads, the bulkhead could terminate at the continuous ceiling or lining.

5.4.2 *Method IF.* All bulkheads not required by this or other sections of this part to be "A" or "B" class divisions should be at least "C" class divisions.

5.4.3 *Method IIF.* There should be no restriction on the construction of bulkheads not required by this or other sections of this part to be "A" or "B" class divisions except in individual cases where "C" class bulkheads are required in accordance with table 1 in 5.7.

5.4.4 *Method IIIF.* There should be no restriction on the construction of bulkheads not required by this or other sections of this part to be "A" or "B" class divisions. In no case should the area of any accommodation space or spaces bounded by a continuous "A" or "B" class division exceed 50 m^2, except in individual cases where "C" class bulkheads are required in accordance with table 1 in 5.7. However, the Competent Authority may increase this area for public spaces.

5.5 Protection of stairways and lift trunks in accommodation spaces, service spaces and control stations

5.5.1 Stairways which penetrate only a single deck should be protected at least at one level by at least "B-0" class divisions and self-closing doors. Lifts which penetrate only a single deck should be enclosed by "A-0" class divisions with steel doors at both levels. Stairways and lift trunks which penetrate more than a single deck should be enclosed by at least "A-0" class divisions and protected by self-closing doors at all levels.

5.5.2 All stairways should be of steel frame construction except where the Competent Authority permits the use of other equivalent material.

5.6 Doors in fire-resistant divisions

5.6.1 Doors should have resistance to fire, as far as practicable, equivalent to the division in which they are fitted. Doors and door frames in "A" class divisions should be constructed of steel. Doors in "B" class divisions should be non-combustible. Doors fitted in boundary bulkheads of machinery spaces of category A should be self-closing and reasonably gas-tight. The Competent Authority could permit the use of combustible materials in doors separating cabins from the individual interior sanitary accommodation, such as showers, if constructed according to method IF.

5.6.2 Doors required to be self-closing should not be fitted with holdback hooks. However, holdback arrangements fitted with remote release fittings of the fail-safe type could be used.

5.6.3 Ventilation openings could be permitted in and under the doors in corridor bulkheads except that such openings should not be permitted in and under stairway enclosure doors. The openings should be provided only in the lower half of a door. Where such opening is in or under a door, the total net area of any such opening or openings should not exceed $0.05\ m^2$. When such opening is cut in a door, it should be fitted with a grille made of non-combustible material.

5.6.4 Watertight doors need not be insulated.

5.7 Fire integrity of bulkheads and decks

5.7.1 In addition to the specific provisions for fire integrity of bulkheads and decks required elsewhere in this part, the minimum fire integrity of bulkheads and decks should be as prescribed in tables 1 and 2 of this section.

5.7.2 The following requirements should govern application of the tables:

.1 tables 1 and 2 should apply respectively to bulkheads and decks separating adjacent spaces; and

.2 for determining the appropriate fire integrity standards to be applied to divisions between adjacent spaces, such spaces are classified according to their fire risk as follows:

Chapter V: part B

(1) *Control stations*
Spaces containing emergency sources of power and lighting.
Wheelhouse and chartroom.
Spaces containing the vessel's radio equipment.
Fire-extinguishing rooms, fire-control rooms and fire recording stations.
Control room for propulsion machinery when located outside the machinery space.
Spaces containing centralized fire alarm equipment.

(2) *Corridors*
Corridors and lobbies.

(3) *Accommodation spaces*
Spaces as defined in 5.2.2 and 5.2.14 excluding corridors.

(4) *Stairways*
Interior stairways, lifts and escalators other than those wholly contained within the machinery spaces and enclosures thereto. In this connection, a stairway which is enclosed only at one level should be regarded as part of the space from which it is not separated by a fire door.

(5) *Service spaces of low fire risk*
Lockers and storerooms having areas of less than 2 m^2 and laundries.

(6) *Machinery spaces of category A*
Spaces as defined in 5.2.12.

(7) *Other machinery spaces*
Spaces as defined in 5.2.11, including fishmeal processing spaces, but excluding machinery spaces of category A.

(8) *Cargo spaces*
All spaces used for cargo, including cargo oil tanks, and trunkways and hatchways to such spaces.

(9) *Service spaces of high fire risk*
Galleys, pantries containing cooking appliances, paint rooms, lamp rooms, lockers and store-rooms having areas of 2 m^2 or more, and workshops other than those forming part of the machinery spaces.

(10) *Open decks*
Open deck spaces and enclosed promenades, spaces for processing fish in the raw state, fish washing spaces and similar spaces containing no fire risk. The air spaces outside superstructures and deckhouses.

The title of each category is intended to be typical rather than restrictive. The number in parenthesis following each category refers to the applicable column or row in the tables.

Table 1 – Fire integrity of bulkheads separating adjacent spaces

Spaces		(1)	(2)	(3)	(4)	(5)	(6)	(7)	(8)	(9)	(10)
Control stations	(1)	A-0[e]	A-0	A-60	A-0	A-15	A-60	A-15	A-60	A-60	*
Corridors	(2)		C	B-0	B-0 A-0[c]	B-0	A-60	A-0	A-0	A-0	*
Accommodation spaces	(3)			C[a,b]	B-0 A-0[c]	B-0	A-60	A-0	A-0	A-0	*
Stairways	(4)				B-0 A-0[c]	B-0 A-0[c]	A-60	A-0	A-0	A-0	*
Service spaces of low fire risk	(5)					C	A-60	A-0	A-0	A-0	*
Machinery spaces of category A	(6)						*	A-0	A-0	A-60	*
Other machinery spaces	(7)							A-0[d]	A-0	A-0	*
Cargo spaces	(8)								*	A-0	*
Service spaces of high fire risk	(9)									A-0[d]	*
Open decks	(10)										–

(See Notes on page 57)

Chapter V: part B

Table 2 – *Fire integrity of decks separating adjacent spaces*

Space below ↓ \ Space above →		(1)	(2)	(3)	(4)	(5)	(6)	(7)	(8)	(9)	(10)
Control stations	(1)	A-0	A-0	A-0	A-0	A-0	A-60	A-0	A-0	A-0	*
Corridors	(2)	A-0	*	*	A-0	*	A-60	A-0	A-0	A-0	*
Accommodation	(3)	A-60	A-0	*	A-0	*	A-60	A-0	A-0	A-0	*
Stairways	(4)	A-0	A-0	A-0	*	A-0	A-60	A-0	A-0	A-0	*
Service spaces of low fire risk	(5)	A-15	A-0	A-0	A-0	*	A-60	A-0	A-0	A-0	*
Machinery spaces of category A	(6)	A-60	A-60	A-60	A-60	A-60	*	A-60	A-30	A-60	*
Other machinery spaces	(7)	A-15	A-0	A-0	A-0	A-0	A-0	*	A-0	A-0	*
Cargo spaces	(8)	A-60	A-0	A-0	A-0	A-0	A-0	A-0	*	A-0	*
Service spaces of high fire risk	(9)	A-60	A-0	A-0	A-0	A-0	A-0	A-0	A-0	A-0^d	*
Open decks	(10)	*	*	*	*	*	*	*	*	*	—

(See Notes on page 57)

Fire protection, detection, extinction and fighting

Notes: To be applied to both tables 1 and 2, as appropriate

(a) No special requirements are imposed upon these bulkheads in methods IIF and IIIF fire protection.

(b) In case of method IIIF, "B" class bulkheads of "B-0" rating should be provided between spaces or groups of spaces of 50 m^2 and over in area.

(c) For clarification as to which applies, see 5.4 and 5.5.

(d) Where spaces are of the same numerical category and superscript ^{d} appears, a bulkhead or deck of the rating shown in the tables is only required when the adjacent spaces are for a different purpose, e.g., in category (9). A galley next to a galley does not require a bulkhead but a galley next to a paint room requires an "A-0" bulkhead.

(e) Bulkheads separating the wheelhouse, chartroom and radio room from each other may be "B-0" rating.

(f) Fire insulation need not be fitted if the machinery space in category (7), in the opinion of the Competent Authority, has little or no fire risk.

* Where an asterisk appears in the tables, the division should be of steel or equivalent material, but does not need to be of "A" class standard. However, where a deck is penetrated for the passage of electrical cables, pipes and vent ducts, such penetrations should be made tight to prevent the passage of flame and smoke.

5.7.3 Continuous "B" class ceilings or linings, in association with the relevant decks or bulkheads, could be accepted as contributing, wholly or in part, to the required insulation and integrity of a division.

5.7.4 Windows and skylights to machinery spaces should be as follows:

.1 where skylights can be opened, they should be capable of being closed from outside the space. Skylights containing glass panels should be fitted with external shutters of steel or other equivalent material permanently attached;

.2 glass or similar materials should not be fitted in machinery space boundaries. This does not preclude the use of wire-reinforced glass for skylights and glass in control rooms within the machinery spaces; and

.3 skylights referred to in 5.7.4.1 should be of wire-reinforced glass.

5.7.5 External boundaries which are required by 5.3.1 to be of steel or equivalent material could be pierced for the fitting of windows and sidescuttles provided that there is no requirement elsewhere in this part for such boundaries to have "A" class integrity. Similarly, in such boundaries, which are not required to have "A" class integrity, doors may be of materials to the satisfaction of the Competent Authority.

5.8 Details of construction

5.8.1 *Method IF.* In accommodation and service spaces and control stations, all linings, draught stops, ceilings and their associated grounds should be of non-combustible materials.

5.8.2 *Methods IIF and IIIF.* In corridors and stairway enclosures serving accommodation and service spaces and control stations, ceilings, linings, draught stops and their associated grounds should be of non-combustible materials.

5.8.3 *Methods IF, IIF and IIIF*

.1 Except in cargo spaces or refrigerated compartments of service spaces, insulating materials should be non-combustible. Vapour barriers and adhesives used in conjunction with insulation, as well as the insulation of pipe fittings, for cold service systems need not be of non-combustible material, but they should be kept to the minimum quantity practicable and their exposed surfaces should have qualities of resistance to the propagation of flame to the satisfaction of the Competent Authority. In spaces where penetration of oil products is possible, the surface of insulation should be impervious to oil or oil vapour.

.2 Where non-combustible bulkheads, linings and ceilings are fitted in accommodation and service spaces, they could have a combustible veneer not exceeding 2 mm in thickness within any such space except corridors, stairway enclosures and control stations, where it should not exceed 1.5 mm in thickness.

.3 Air spaces enclosed behind ceilings, panellings or linings should be divided by close-fitting draught stops spaced not more than 14 m apart. In the vertical direction, such spaces, including those behind linings of stairways, trunks, etc., should be closed at each deck.

5.9 Ventilation systems

5.9.1 Ventilation ducts should be of non-combustible material. Short ducts, however, not generally exceeding 2 m in length and with a cross-section not exceeding 0.02 m^2 need not be non-combustible, subject to the following conditions:

.1 these ducts should be of a material which, to the satisfaction of the Competent Authority, has a low fire risk;

.2 they should only be used at the end of the ventilation device; and

.3 they should not be situated less than 600 mm, measured along the duct, from an opening in an "A" or "B" class division, including continuous "B" class ceilings.

Fire protection, detection, extinction and fighting

5.9.2 Where the ventilation ducts with a free cross-sectional area exceeding 0.02 m² pass through "A" class bulkheads or decks, the opening should be lined with a steel sheet sleeve unless the ducts passing through the bulkheads or decks are of steel in the vicinity of passage through the deck or bulkhead and comply in that portion of the duct with the following:

- .1 for ducts with a free cross-sectional area exceeding 0.02 m², the sleeves should have a thickness of at least 3 mm and a length of at least 900 mm. When passing through bulkheads, this length should preferably be divided evenly on each side of the bulkhead. Ducts with free cross-sectional area exceeding 0.02 m² should be provided with fire insulation. The insulation should have at least the same fire integrity as the bulkhead or deck through which the duct passes. Equivalent penetration protection should be provided to the satisfaction of the Competent Authority; and

- .2 ducts with a free cross-sectional area exceeding 0.085 m² should be fitted with fire dampers in addition to the recommendations of 5.9.2.1. The fire damper should operate automatically but should also be capable of being closed manually from both sides of the bulkhead or deck. The damper should be provided with an indicator which shows whether the damper is open or closed. Fire dampers are not required, however, where ducts pass through spaces surrounded by "A" class divisions without serving those spaces, provided those ducts have the same fire integrity as the bulkheads which they penetrate.

5.9.3 Ventilation ducts for machinery spaces of category A or galleys should not, in general, pass through accommodation spaces, service spaces or control stations. Where the Competent Authority permits this arrangement, the ducts should be constructed of steel or equivalent material and so arranged as to preserve the integrity of the divisions.

5.9.4 Ventilation ducts of accommodation spaces, service spaces or control stations should not, in general, pass through machinery spaces of category A or through galleys. Where the Competent Authority permits this arrangement, the ducts should be constructed of steel or equivalent material and so arranged as to preserve the integrity of the divisions.

5.9.5 Where ventilation ducts with a free cross-sectional area exceeding 0.02 m² pass through "B" class bulkheads, the openings should be lined with steel sheet sleeves of at least 900 mm in length, unless the ducts are of steel for this length in way of the bulkheads. When passing through a "B" class bulkhead, this length should preferably be divided evenly on each side of the bulkhead.

5.9.6 Practical measures should be taken in respect of control stations outside machinery spaces in order to ensure that ventilation, visibility and freedom from smoke are maintained, so that in the event of fire the machinery and equipment contained therein may be supervised and

Chapter V: part B

continue to function effectively. Alternative and separate means of air supply should be provided; air inlets of the two sources of supply should be so disposed that the risk of both inlets drawing in smoke simultaneously is minimized. At the discretion of the Competent Authority, such requirements need not apply to control stations situated on, and opening onto, an open deck, or where local closing arrangements are equally effective.

5.9.7 Where they pass through accommodation spaces or spaces containing combustible materials, the exhaust ducts from galley ranges should be constructed of "A" class divisions. Each exhaust duct should be fitted with:

    .1   a grease trap readily removable for cleaning;

    .2   a fire damper located in the lower end of the duct;

    .3   arrangements, operable from within the galley, for shutting off the exhaust fan; and

    .4   fixed means for extinguishing a fire within the duct, except where the Competent Authority considers such fittings impractical in a vessel of less than 75 m in length.

5.9.8 The main inlets and outlets of all ventilation systems should be capable of being closed from outside the spaces being ventilated. Power ventilation of accommodation spaces, service spaces, control stations and machinery spaces should be capable of being stopped from an easily accessible position outside the space being served. This position should not be readily cut off in the event of a fire in the spaces served. The means provided for stopping the power ventilation of the machinery spaces should be entirely separate from the means provided for stopping ventilation of other spaces.

5.9.9 Means should be provided for closing, from a safe position, the spaces around funnels.

5.9.10 Ventilation systems serving machinery spaces should be independent of systems serving other spaces.

5.9.11 Store-rooms containing appreciable quantities of highly flammable products should be provided with ventilation arrangements which are separate from other ventilation systems. Ventilation should be arranged at high and low levels and the inlets and outlets of ventilators should be positioned in safe areas and fitted with spark arresters.

5.10   Heating installations

5.10.1 Electric radiators should be fixed in position and so constructed as to reduce fire risks to a minimum. No such radiator should be fitted with an element so exposed that clothing, curtains, or other similar materials can be scorched or set on fire by heat from the element.

5.10.2 Heating by means of open fires should not be permitted. Heating stoves and other similar appliances should be firmly secured and adequate protection and insulation against fire should be provided beneath and around such appliances and in way of their uptakes. Uptakes of stoves which burn solid fuel should be so arranged and designed as to minimize the possibility of becoming blocked by combustion products and should have a ready means for cleaning. Dampers for limiting draughts in uptakes should, when in the closed position, still leave an adequate area open. Spaces in which stoves are installed should be provided with ventilators of sufficient area to provide adequate combustion-air for the stove. Such ventilators should have no means of closure and their position should be such that closing appliances in accordance with 2.9 are not required.

5.10.3 Open-flame gas appliances, except cooking stoves and water heaters, should not be permitted. Spaces containing any such stoves or water heaters should have adequate ventilation to remove fumes and possible gas leakage to a safe place. All pipes conveying gas from container to stove or water heater should be of steel or other approved material. Automatic safety gas shut-off devices should be fitted to operate on loss of pressure in the gas main pipe or flame failure on any appliance.

5.10.4 Where gaseous fuel is used for domestic purposes, the arrangements, storage, distribution and use of the fuel should be to the satisfaction of the Competent Authority and in accordance with 5.12.

5.11 Miscellaneous items*

5.11.1 All exposed surfaces in corridors and stairway enclosures and surfaces including grounds in concealed or inaccessible spaces in accommodation and service spaces and control stations should have low flame-spread characteristics.† Exposed surfaces of ceilings in accommodation and service spaces and control stations should have low flame-spread characteristics.

5.11.2 Paints, varnishes and other finishes used on exposed interior surfaces should not be capable of producing excessive quantities of smoke or toxic gases or vapours, to be determined in accordance with the Fire Test Procedures Code.

5.11.3 Primary deck coverings within accommodation and service spaces and control stations should be of approved material which will not readily

---

* Refer to the Guidance concerning the use of certain plastic materials, contained in Recommendation 7 of attachment 3 to the Final Act of the 1993 Torremolinos Conference.

† Refer to the Guidelines for the evaluation of fire hazard properties of materials, adopted by the Organization by resolution A.166(ES.IV), and the Recommendation on improved fire test procedures for surface flammability of bulkhead, ceiling and deck finish materials, adopted by the Organization by resolution A.653(16).

ignite or give rise to toxic or explosive hazards at elevated temperatures, this being determined in accordance with the Fire Test Procedures Code.*

5.11.4 Where "A" or "B" class divisions are penetrated for the passage of electrical cables, pipes, trunks, ducts, etc., or for the fitting of ventilation terminals, lighting fixtures and similar devices, arrangements should be made to ensure that the fire integrity of the divisions is not impaired.

5.11.5 In accommodation and service spaces and control stations, pipes penetrating "A" or "B" class divisions should be of approved materials having regard to the temperature that such divisions are required to withstand. Where the Competent Authority permits the conveying of oil and combustible liquids through accommodation and service spaces, the pipes conveying oil or combustible liquids should be of an approved material having regard to the fire risk.

5.11.6 Materials readily rendered ineffective by heat should not be used for overboard scuppers, sanitary discharges, and other outlets which are close to the waterline and where the failure of the material in the event of fire would give rise to danger of flooding.

5.11.7 Cellulose-nitrate-based film should not be used in cinematography installations.

5.11.8 All waste receptacles other than those used in fish processing should be constructed of non-combustible materials with no openings in the sides or bottom.

5.11.9 Machinery driving fuel oil transfer pumps, fuel oil unit pumps and other similar fuel pumps should be fitted with remote controls situated outside the space concerned, so that they can be stopped in the event of a fire arising in the space in which they are located.

5.11.10 Drip trays should be fitted, where necessary, to prevent oil leaking into bilges.

5.11.11 Within compartments used for stowage of fish, combustible insulation should be protected by close-fitting cladding.

5.12 Storage of gas cylinders and dangerous materials

5.12.1 Cylinders for compressed, liquefied, or dissolved gases should be clearly marked by means of prescribed identifying colours; have a clearly legible identification of the name and chemical formula of their contents; and should be properly secured.

5.12.2 Cylinders containing flammable or other dangerous gases and empty cylinders should be stored, properly secured, on open decks and all valves, pressure regulators and pipes leading from such cylinders should

---

* Refer to the Recommendation on fire test procedures for ignitability of primary deck coverings, adopted by the Organization by resolution A.687(17).

be protected against damage. Cylinders should be protected against excessive variations in temperature, direct rays of the sun and accumulation of snow. However, the Competent Authority may permit such cylinders to be stored in compartments complying with the recommendations of 5.12.3 to 5.12.5.

5.12.3 Spaces containing highly flammable liquids, such as volatile paints, paraffin, benzole, etc. and, where permitted, liquefied gas, should have direct access from open decks only. Pressure-adjusting devices and relief valves should exhaust within the compartment. Where boundary bulkheads of such compartments adjoin other enclosed spaces, they should be gas-tight.

5.12.4 Except as necessary for service within the space, electrical wiring and fittings should not be permitted within compartments used for the storage of highly flammable liquids or liquefied gases. Where such electrical fittings are installed, they should be to the satisfaction of the Competent Authority for use in a flammable atmosphere. Sources of heat should be kept clear of such spaces and "No Smoking" and "No Naked Light" notices should be displayed in a prominent position.

5.12.5 Separate storage should be provided for each type of compressed gas. Compartments used for the storage of such gases should not be used for storage of other combustible products nor for tools or objects not part of the gas distribution system. However, the Competent Authority may relax these recommendations considering the characteristics, volume and intended use of such compressed gases.

5.13 Means of escape

5.13.1 Stairways and ladders leading to and from all accommodation spaces and in spaces in which the crew is normally employed, other than machinery spaces, should be so arranged as to provide ready means of escape to the open deck and thence to the survival craft. In particular, in relation to these spaces:

.1 at all levels of accommodation, at least two widely separated means of escape should be provided which may include the normal means of access from each restricted space or group of spaces;

.2.1 below the weather deck, the main means of escape should be a stairway and the second escape may be a trunk or a stairway; and

.2.2 above the weather deck, the means of escape should be stairways or doors to an open deck or a combination thereof;

.3 exceptionally, the Competent Authority may permit only one means of escape, due regard being paid to the nature and location of spaces and to the number of persons who normally might be accommodated or employed there;

> .4 a corridor or part of a corridor from which there is only one route of escape should not exceed 7 m in length; and
>
> .5 the width and continuity of the means of escape should be to the satisfaction of the Competent Authority.

5.13.2 Two means of escape should be provided from every machinery space of category A by one of the following means:

> .1 two sets of steel ladders, as widely separated as possible, leading to doors in the upper part of the space similarly separated and from which access is provided to the open deck. In general, one of these ladders should provide continuous fire shelter from the lower part of the space to a safe position outside the space. However, the Competent Authority may not require such shelter if, due to special arrangements or dimensions of the machinery space, a safe escape route from the lower part of this space is provided. This shelter should be of steel, insulated and be provided with a self-closing steel door at the lower end; or
>
> .2 one steel ladder leading to a door in the upper part of the space from which access is provided to the open deck and additionally, in the lower part of the space and in a position well separated from the ladder referred to, a steel door capable of being operated from each side and which provides access to a safe escape route from the lower part of the space to the open deck.

5.13.3 From machinery spaces other than those of category A, escape routes should be provided to the satisfaction of the Competent Authority having regard to the nature and location of the space and whether persons are normally employed in that space.

5.13.4 Lifts should not be considered as forming one of the required means of escape.

5.14 Automatic sprinkler and fire alarm and fire detection systems (method IIF)

5.14.1 In vessels in which method IIF is adopted, an automatic sprinkler and fire alarm system of an approved type and complying with the recommendations of this section should be installed and so arranged as to protect accommodation spaces and service spaces except spaces which afford no substantial fire risks, such as void spaces and sanitary spaces.

5.14.2 The system should be capable of immediate operation at all times and no action by the crew should be necessary to set it in operation. It should be of the wet-pipe type but small exposed sections could be of the dry-pipe type where, in the opinion of the Competent Authority, this is a necessary precaution. Any parts of the system which may be subjected to

Fire protection, detection, extinction and fighting

freezing temperatures in service should be suitably protected against freezing.* It should be kept charged at the necessary pressure and should have provision for a continuous supply of water as required in 5.14.13.

5.14.3 Each section of sprinklers should include means for giving a visible and audible alarm signal automatically at one or more indicating units whenever any sprinkler comes into operation. Such units should indicate in which section served by the system, fire has occurred and should be centralized in the wheelhouse and, in addition, visible and audible alarms from the unit should be placed in a position other than in the wheelhouse, so as to ensure that the indication of fire is immediately received by the crew. Such an alarm system should be so constructed as to indicate if any fault occurs in the system.

5.14.4 Sprinklers should be grouped into separate sections, each of which should contain not more than 200 sprinklers.

5.14.5 Each section of sprinklers should be capable of being isolated by one stop valve only. The stop valve in each section should be readily accessible and its location should be clearly and permanently indicated. Means should be provided to prevent the operation of the stop valves by any unauthorized person.

5.14.6 A gauge indicating the pressure in the system should be provided at each section stop valve and at a central station.

5.14.7 The sprinklers should be resistant to corrosion. In accommodation and service spaces, the sprinklers should come into operation within the temperature range of 68°C and 79°C, except that in locations such as drying rooms, where high ambient temperatures might be expected, the operating temperature could be increased by not more than 30°C above the maximum deckhead temperature.

5.14.8 A list or plan should be displayed at each indicating unit showing the spaces covered and the location of the zone in respect of each section. Suitable instructions for testing and maintenance should be available.

5.14.9 Sprinklers should be placed in an overhead position and spaced in a suitable pattern to maintain an average application rate of not less than 5 l per m^2 per minute over the nominal area covered by the sprinklers. Alternatively, the Competent Authority may permit the use of sprinklers providing such quantity of water suitably distributed as has been shown to the satisfaction of the Competent Authority to be not less effective.

5.14.10 A pressure tank having a volume equal to at least twice that of the charge of water specified in this subparagraph should be provided. The tank should contain a standing charge of fresh water, equivalent to the amount of water which would be discharged in one minute by the pump

---

* Refer to the Guidance on precautions against freezing of fire mains, contained in recommendation 6 of the attachment to the Final Act of the 1993 Torremolinos Conference.

referred to in 5.14.13, and the arrangements should provide for maintaining such air pressure in the tank as to ensure that, where the standing charge of fresh water in the tank has been used, the pressure will be not less than the working pressure of the sprinkler plus the pressure due to a head of water measured from the bottom of the tank to the highest sprinkler in the system. Suitable means of replenishing the air under pressure and of replenishing the fresh water charge in the tank should be provided. A glass gauge should be provided to indicate the correct level of the water in the tank.

5.14.11 Means should be provided to prevent the passage of seawater into the tank.

5.14.12 An independent power pump should be provided solely for the purpose of continuing automatically the discharge of water from the sprinklers. The pump should be brought into action automatically by the pressure drop in the system before the standing fresh water charge in the pressure tank is completely exhausted.

5.14.13 The pump and the piping system should be capable of maintaining the necessary pressure at the level of the highest sprinkler to ensure a continuous output of water sufficient for the simultaneous coverage of the maximum area separated by fire-resisting bulkheads of "A" and "B" class divisions or an area of 280 m^2, whichever is the less, at the application rate specified in 5.14.9.

5.14.14 The pump should be fitted, on the delivery side, with a test valve, with a short open-ended discharge pipe. The effective area through the valve and pipe should be adequate to permit the release of the required pump output while maintaining the pressure in the system specified in 5.14.10.

5.14.15 The sea inlet to the pump should, wherever possible, be in the space containing the pump and should be so arranged that, when the vessel is afloat, it will not be necessary to shut off the supply of seawater to the pump for any purpose other than the inspection or repair of the pump.

5.14.16 The sprinkler pump and tank should be situated in a position reasonably remote from any machinery space of category A and should not be situated in any space required to be protected by the sprinkler system.

5.14.17 There should not be less than two sources of power supply for the seawater pump and the automatic fire alarm and fire detection system. If the pump is electrically driven it should be connected to the main source of electrical power, which should be capable of being supplied by at least two generators.

5.14.18 The feeders should be arranged so as to avoid galleys, machinery spaces and other enclosed spaces of high fire risk, except insofar as it is necessary to reach the appropriate switchboard. One of the sources of power supply for the fire alarm and fire detection system should be an emergency source. Where one of the sources of power for the pump is an internal-combustion-type engine, it should, in addition to complying with the provisions of 5.14.16, be so situated that a fire in any protected space will not affect the air supply to that engine.

5.14.19 The sprinkler system should have a connection from the vessel's fire main by way of a lockable screw-down non-return valve at the connection, which will prevent a backflow from the sprinkler system to the fire main.

5.14.20 A test valve should be provided for testing the automatic alarm for each section of sprinklers by a discharge of water equivalent to the operation of one sprinkler. The test valve for each section should be situated near the stop valve for that section.

5.14.21 Means should be provided for testing the automatic operation of the pump on reduction of pressure in the system.

5.14.22 Switches should be provided at one of the indicating positions referred to in 5.14.3, which will enable the alarm and the indicators for each section of sprinklers to be tested.

5.14.23 Spare sprinkler heads should be provided for each section of sprinklers to the satisfaction of the Competent Authority.

5.15 Automatic fire alarm and fire detection systems (method IIIF)

5.15.1 In vessels in which method IIIF is adopted, an automatic fire alarm and fire detection system of an approved type and complying with the recommendations of this section should be installed and so arranged as to detect the presence of fire in all accommodation spaces and service spaces except spaces which afford no substantial fire risk, such as void spaces and sanitary spaces.

5.15.2 The system should be capable of immediate operation at all times and no action of the crew should be necessary to set it in operation.

5.15.3 Each section of detectors should include means for giving a visible and audible alarm signal automatically at one or more indicating units whenever any detector comes into operation. Such units should indicate in which section served by the system a fire has occurred and should be centralized on the wheelhouse and such other positions as will ensure that any alarm from the system is immediately received by the crew. Additionally, arrangements should be provided to ensure that an alarm is sounded on the deck on which the fire has been detected. Such an alarm and detection system should be so constructed as to indicate if any fault occurs in the system.

5.15.4 Detectors should be grouped into separate sections, each covering not more than 50 rooms served by such a system and containing not more than 100 detectors. Detectors should be zoned to indicate on which deck a fire has occurred.

5.15.5 The system should be operated by an abnormal air temperature, by an abnormal concentration of smoke or by other factors indicative of incipient fire in any one of the spaces to be protected. Systems which are sensitive to air temperature should not operate at less than 54°C and should operate at a temperature not greater than 78°C when the temperature increase to those levels is not more than 1°C per minute. At the discretion of the Competent Authority, the permissible temperature of operation could be increased to 30°C above the maximum deckhead temperature in drying rooms and similar places of a normally high ambient temperature. Systems which are sensitive to smoke concentration should operate on the reduction of the intensity of a transmitted light beam by an amount to be determined by the Competent Authority. Other equally effective methods of operation could be accepted at the discretion of the Competent Authority. The detection system should not be used for any purpose other than fire detection.

5.15.6 The detectors should be arranged to operate the alarm by the opening or closing of contacts or by other appropriate methods. They should be fitted in an overhead position and should be suitably protected against impact and physical damage. They should be suitable for use in a marine atmosphere. They should be placed in an open position clear of beams and other objects likely to obstruct the flow of hot gases or smoke to the sensitive element. Detectors operated by the closing of contacts should be of the sealed-contact type and the circuit should be continuously monitored to indicate fault conditions.

5.15.7 At least one detector should be installed in each space where detection facilities are required and there should be not less than one detector for each 37 m^2 of deck area approximately. In large spaces, the detectors should be arranged in a regular pattern so that no detector is more than 9 m from another detector or more than 4.5 m from a bulkhead.

5.15.8 There should be not less than two sources of power supply for the electrical equipment used in the operation of the fire alarm and fire detection system, one of which should be an emergency source. The supply should be provided by separate feeders reserved solely for that purpose. Such feeders should run to a changeover switch situated in the control station for the fire detection system. The wiring system should be so arranged as to avoid galleys, machinery spaces and other enclosed spaces having a high fire risk except, insofar as it is necessary, to provide for fire detection in such spaces or to reach the appropriate switchboard.

5.15.9 A list or plan should be displayed adjacent to each indicating unit showing the spaces covered and the location of the zone in respect of each system. Suitable instructions for testing and maintenance should be available.

Fire protection, detection, extinction and fighting

5.15.10 Provision should be made for testing the correct operation of the detectors and the indicating units by supplying means for applying hot air or smoke at detector positions.

5.15.11 Spare detector heads should be provided for each section of detectors to the satisfaction of the Competent Authority.

5.16 Fixed fire-extinguishing arrangements in cargo spaces of high fire risk

Cargo spaces of high fire risk should be protected by a fixed gas fire-extinguishing system complying with the Fire Safety System Code or by a fire-extinguishing system which gives equivalent protection, to the satisfaction of the Competent Authority.

5.17 Fire pumps

5.17.1 At least two fire pumps should be provided.

5.17.2 If a fire in any one compartment could put all the fire pumps out of action, there should be an alternative means of providing water for fire fighting. In vessels of 75 m in length and over, this alternative means should be a fixed emergency fire pump independently driven. This emergency fire pump should be capable of supplying two jets of water to the satisfaction of the Competent Authority.

5.17.3 The fire pumps, other than the emergency pump, should be capable of delivering for fire-fighting purposes a quantity of water at a minimum pressure of 0.25 N/mm², with a total capacity (Q) of at least:

$$Q = (0.15 \sqrt{L(B+D)} + 2.25)^2 \text{ m}^3/\text{h}$$

where L, B and D are in metres.

However, the total required capacity of the fire pumps need not exceed 180 m³/h.

5.17.4 Each of the required fire pumps other than any emergency pump should have a capacity not less than 40% of the total capacity of fire pumps required by 5.17.3 and should, in any event, be capable of delivering at least the jets of water required by 5.19.3. These fire pumps should be capable of supplying the fire main systems under the required conditions. Where more than two pumps are installed, the capacity of such additional pumps should be to the satisfaction of the Competent Authority.

5.17.5 Fire pumps should be independently driven power pumps. Sanitary, ballast, bilge or general service pumps can be accepted as fire pumps, provided that they are not normally used for pumping oil and that, if they are subject to occasional duty for the transfer or pumping of fuel oil, suitable change-over arrangements are fitted.

5.17.6 Relief valves should be provided in conjunction with all fire pumps if the pumps are capable of developing a pressure exceeding the design pressure of the water service pipes, hydrants and hoses. These valves should be so placed and adjusted as to prevent excessive pressure in any of the fire main systems.

5.17.7 Emergency power-operated fire pumps should be independently driven self-contained pumps either with their own diesel engine prime mover and fuel supply fitted in an accessible position outside the compartment which contains the main fire pumps, or be driven by a self-contained generator, which can be the emergency generator referred to in 4.16, of sufficient capacity and which is positioned in a safe place outside the engine-room and preferably above the working deck. The emergency fire pump should be capable of operating for a period of at least 3 h.

5.17.8 Emergency fire pumps, sea-suction valves and other necessary valves should be operable from outside compartments containing main fire pumps in a position not likely to be cut off by a fire in those compartments.

5.18 Fire mains

5.18.1 Where more than one hydrant is required to provide the number of jets specified in 5.19.3, a fire main should be provided.

5.18.2 Fire mains should have no connections other than those required for fire fighting, except for the purpose of washing the deck and anchor chains and operation of bilge ejectors, subject to the efficiency of the fire-fighting system being maintained.

5.18.3 Where fire mains are not self-draining, suitable drain cocks should be fitted where frost damage could be expected.*

5.18.4 The diameter of the fire main and water service pipes should be sufficient for the effective distribution of the maximum required discharge from two fire pumps operating simultaneously or of 140 m^3 per hour, whichever is the less.

5.18.5 With the two pumps simultaneously delivering through nozzles specified in 5.19.7 the quantity of water specified in 5.18.4, through any adjacent hydrants, the minimum pressure of 0.25 N/mm^2 should be maintained at all hydrants.

5.19 Fire hydrants, fire hoses and nozzles

5.19.1 The number of fire hoses provided should be equal to the number of fire hydrants arranged according to 5.19.3 and one spare hose. This

---

* Refer to the Guidance on precautions against freezing of fire mains, contained in recommendation 6 of the attachment to the Final Act of the 1993 Torremolinos Conference.

number does not include any fire hoses required in any engine or boiler room. The Competent Authority can increase the number of fire hoses required so as to ensure that hoses in sufficient number are available and accessible at all times, having regard to the size of the vessel.

5.19.2 Fire hoses should be of approved material and sufficient in length to project a jet of water to any of the spaces in which they may be required to be used. Their maximum length should be 20 m. Every fire hose should be provided with a nozzle and the necessary couplings. Fire hoses should, together with any necessary fittings and tools, be kept ready for use in conspicuous positions near the water service hydrants or connections.

5.19.3 The number and position of the hydrants should be such that at least two jets of water not emanating from the same hydrant, one of which should be from a single length of fire hose, may reach any part of the vessel normally accessible to the crew while the vessel is being navigated.

5.19.4 All required hydrants should be fitted with fire hoses having dual-purpose nozzles as required by 5.19.7. One hydrant should be located near the entrance of the space to be protected.

5.19.5 Materials readily rendered ineffective by heat should not be used for fire mains and hydrants unless adequately protected. The pipes and hydrants should be so placed that the fire hoses may be easily coupled to them. In vessels where deck cargo can be carried, the positions of the hydrants should be such that they are always readily accessible and the pipes should be arranged, as far as practicable, to avoid risk of damage by such cargo. Unless one fire hose and nozzle is provided for each hydrant, there should be complete interchangeability of fire hose couplings and nozzles.

5.19.6 A cock or valve should be fitted to serve each fire hose so that any fire hose may be removed while the fire pumps are operating.

5.19.7 Standard nozzle sizes should be 12 mm, 16 mm and 19 mm or as near to as possible. Larger diameter nozzles can be permitted at the discretion of the Competent Authority.

5.19.8 For accommodation and service spaces, a nozzle size greater than 12 mm need not be used.

5.19.9 For machinery spaces and exterior locations, the nozzle size should be such as to obtain the maximum discharge possible from two jets at the pressure specified in 5.18.5 from the smallest pump, provided that a nozzle size greater than 19 mm need not be used.

5.20 Fire extinguishers*

5.20.1 Fire extinguishers should be of approved types. The capacity of required portable fluid extinguishers should be not more than 13.5 l and not less than 9 l. Other extinguishers should not be in excess of the equivalent portability of the 13.5 l fluid extinguisher and should not be less than the fire-extinguishing equivalent of a 9 l fluid extinguisher. The Competent Authority should determine the equivalents of fire extinguishers.

5.20.2 Spare charges should be provided to the satisfaction of the Competent Authority.

5.20.3 Fire extinguishers containing an extinguishing medium which, in the opinion of the Competent Authority, either by itself or under expected conditions of use, gives off toxic gases in such quantities as to endanger persons should not be permitted.

5.20.4 Fire extinguishers should be periodically examined and subject to such tests as the Competent Authority may require.

5.20.5 Normally, one of the portable fire extinguishers intended for use in any space should be stowed near an entrance to that space.

5.21 Portable fire extinguishers in control stations and accommodation and service spaces

5.21.1 At least five approved portable fire extinguishers should be provided in control stations and accommodation and service spaces to the satisfaction of the Competent Authority.

5.21.2 Spare charges should be provided to the satisfaction of the Competent Authority.

5.22 Fire-extinguishing appliances in machinery spaces

5.22.1 Spaces containing oil-fired boilers or fuel oil units should be provided with one of the following fixed fire-extinguishing systems, to the satisfaction of the Competent Authority:

    .1   a pressure water-spraying installation;

    .2   a fire-smothering gas installation;

---

* Refer to the Improved Guidelines for marine portable fire extinguishers, adopted by the Organization by resolution A.951(23).

Fire protection, detection, extinction and fighting

    .3   a fire-extinguishing installation using vapours from low-toxicity vaporizing liquids; or

    .4   a fire-extinguishing installation using high-expansion foam.

Where the engine and boiler rooms are not entirely separate, or if fuel oil can drain from the boiler room into the engine-room, the combined engine and boiler rooms should be considered as one compartment.

5.22.2 New installations of halogenated hydrocarbon systems used as fire-extinguishing media should be prohibited on new and existing vessels.

5.22.3 Every boiler room should be provided with at least one set of portable air-foam equipment to the satisfaction of the Competent Authority.

5.22.4 At least two approved portable extinguishers discharging foam or equivalent should be provided in each firing space in each boiler room and each space in which a part of the fuel oil installation is situated. At least one approved foam-type extinguisher of at least 135 l capacity or equivalent should be provided in each boiler room. These extinguishers should be provided with hoses on reels suitable for reaching any part of the boiler room. The Competent Authority can relax the recommendations of this paragraph, having regard to the size and nature of the space to be protected.

5.22.5 In each firing space, there should be a receptacle containing sand, sawdust impregnated with soda or other approved dry material, in such quantity as may be required by the Competent Authority. Alternatively, an approved portable extinguisher may be provided.

5.22.6 Spaces containing internal-combustion machinery used either for main propulsion or for other purposes, when such machinery has a total power output of not less than 750 kW, should be provided with the following arrangements:

    .1   one of the fire-extinguishing systems required by 5.22.1;

    .2   at least one set of portable air-foam equipment to the satisfaction of the Competent Authority; and

    .3   in each such space, approved foam-type fire extinguishers each of at least 45 l capacity, or equivalent, sufficient in number to enable foam or its equivalent to be directed onto any part of the fuel and lubricating oil pressure systems, gearing and other fire hazards. In addition, there should be provided a sufficient number of portable foam extinguishers or equivalent, which should be so located that an extinguisher is not more than 10 m walking distance from any point in the space, provided that there should be at least two such extinguishers in each such space. For smaller spaces, the Competent Authority can relax these recommendations.

Chapter V: part B

5.22.7 Spaces containing steam turbines or enclosed steam engines used either for main propulsion or for other purposes, when such machinery has a total power output of not less than 750 kW, should be provided with the following arrangements:

.1 foam fire extinguishers, each of at least 45 l capacity, or equivalent, sufficient in number to enable foam or its equivalent to be directed onto any part of the pressure lubrication system, onto any part of the casings enclosing pressure-lubricated parts of the turbines, engines or associated gearing, and any other fire hazards. Such extinguishers need not be required if such spaces are protected by a fixed fire-extinguishing system fitted in compliance with 5.22.1; and

.2 a sufficient number of portable foam extinguishers, or equivalent, which should be so located that an extinguisher is not more than 10 m walking distance from any point in the space; provided that there should be at least two such extinguishers in each such space, and such extinguishers should not be required in addition to any provided in compliance with 5.22.6.3.

5.22.8 Where, in the opinion of the Competent Authority, a fire hazard exists in any machinery space for which no specific provisions for fire-extinguishing appliances are prescribed in 5.22.1, 5.22.6 and 5.22.7, there should be provided in, or adjacent to, that space a number of approved portable fire extinguishers or other means of fire extinction to the satisfaction of the Competent Authority.

5.22.9 Where fixed fire-extinguishing systems not required by this part are installed, such systems should be to the satisfaction of the Competent Authority.

5.22.10 For any machinery space of category A to which access is provided at a low level from an adjacent shaft tunnel, there should be provided, in addition to any watertight door and on the side remote from that machinery space, a light steel fire-screen door which should be capable of being operated from each side of the door.

5.23 International shore connection

5.23.1 At least one international shore connection complying with 5.23.2 should be provided.

5.23.2 Standard dimensions of flanges for the international shore connection should be in accordance with the following table:

Description	Dimension
Outside diameter	178 mm
Inner diameter	64 mm
Bolt circle diameter	132 mm
Slots in flange	4 holes 19 mm in diameter equidistantly placed on a bolt circle of the above diameter, slotted to the flange periphery
Flange thickness	14.5 mm minimum
Bolts and nuts	4 each of 16 mm in diameter and 50 mm in length

5.23.3 This connection should be constructed of material suitable for 1 N/mm^2 service pressure.

5.23.4 The flange should have a flat face on one side and the other should have a coupling permanently attached thereto that will fit the vessel's hydrant and hose. The connection should be kept aboard the vessel together with a gasket of any material suitable for 1 N/mm^2 service pressure together with four 16 mm bolts 50 mm in length and eight washers.

5.23.5 Facilities should be available enabling such a connection to be used on either side of the vessel.

5.24 Firefighters' outfits

5.24.1 At least two firefighters' outfits should be carried. The firefighters' outfits should be in accordance with paragraphs 2.1, 2.1.1, 2.1.2 and 2.1.3 of chapter 3 of the Fire Safety Systems Code.

5.24.2 The firefighters' outfits should be stored so as to be easily accessible and ready for use and should be stored in widely separated positions.

5.25 Fire control plan

There should be a permanently exhibited fire control plan to the satisfaction of the Competent Authority.*

---

* Refer to the Graphical symbols for shipboard fire control plans, adopted by the Organization by resolution A.952(23).

5.26 Ready availability of fire-extinguishing appliances

Fire-extinguishing appliances should be kept in good order and available for immediate use at all times.

5.27 Acceptance of substitutes

Where in this part any special type of appliance, apparatus, extinguishing medium or arrangement is specified, any other type of appliance, etc., can be allowed, provided the Competent Authority is satisfied that it is not less effective.

Part C
Fire safety measures in vessels of 45 m in length and over but less than 60 m

5.28 Structural fire protection

5.28.1 The hull, superstructure, structural bulkheads, decks and deckhouses should be constructed of non-combustible materials. The Competent Authority may permit combustible construction provided the requirements of this section and the additional fire-extinguishing requirements of 5.40.5 are complied with.

5.28.2 In vessels the hull of which is constructed of non-combustible materials, the decks and bulkheads separating machinery spaces of category A from accommodation spaces, service spaces or control stations should be constructed to "A-60" class standard where the machinery space of category A is not provided with a fixed fire-extinguishing system and to "A-30" class standard where such a system is fitted. Decks and bulkheads separating other machinery spaces from accommodation, service spaces and control stations should be constructed to "A-0" class standard. Decks and bulkheads separating control stations from accommodation and service spaces should be constructed to "A" class standard, insulated to the satisfaction of the Competent Authority, except that a Competent Authority may permit the fitting of "B-15" class divisions for separating such spaces as skipper's cabin from the wheelhouse.

5.28.3 In vessels the hull of which is constructed of combustible materials, the decks and bulkheads separating machinery spaces from accommodation spaces, service spaces or control stations should be constructed to "F" class or "B-15" class standard. In addition, machinery space boundaries should, as far as practicable, prevent the passage of smoke. Decks and bulkheads separating control stations from accommodation and service spaces should be constructed to "F" class standard.

Fire protection, detection, extinction and fighting

5.28.4 In vessels the hull of which is constructed of non-combustible materials, bulkheads of corridors serving accommodation spaces, service spaces and control stations should be of "B-15" class divisions.

5.28.5 In vessels the hull of which is constructed of combustible materials, bulkheads of corridors serving accommodation spaces, service spaces and control stations should be of "F" class divisions.

5.28.6 Any bulkhead required by 5.28.4 or 5.28.5 should extend from deck to deck unless a continuous ceiling of the same class as the bulkhead is fitted on both sides of the bulkhead, in which case the bulkhead can terminate at the continuous ceiling.

5.28.7 Interior stairways serving accommodation spaces, service spaces or control stations should be of steel or other equivalent material. Such stairways should be within enclosures constructed of "F" class divisions in vessels the hull of which is constructed of combustible materials, or "B-15" class divisions in vessels the hull of which is constructed of non-combustible materials, provided that where a stairway penetrates only one deck it need be enclosed at one level only.

5.28.8 Doors and other closures of openings in bulkheads and decks referred to in 5.28.2 to 5.28.6, doors fitted to stairway enclosures referred to in 5.28.7 and doors fitted in engine and boiler casings should be, as far as practicable, equivalent in resisting fire to the divisions in which they are fitted. Doors to machinery spaces of category A should be self-closing.

5.28.9 Lift trunks which pass through the accommodation and service spaces should be constructed of steel or equivalent material and should be provided with means of closing which will permit control of draught and smoke.

5.28.10 In vessels the hull of which is constructed of combustible materials, the boundary bulkheads and decks of spaces containing any emergency source of power and bulkheads and decks between galleys, paint rooms, lamp rooms or any store-rooms which contain appreciable quantities of highly flammable materials and accommodation spaces, service spaces or control stations should be constructed of "F" class or "B-15" class divisions.

5.28.11 In vessels the hull of which is constructed of non-combustible materials, the decks and bulkheads referred to in 5.28.10 should be of "A" class divisions insulated to the satisfaction of the Competent Authority, having in mind the risk of fire, except that the Competent Authority can accept "B-15" class divisions between galley and accommodation spaces, service spaces and control stations when the galley contains electrically heated furnaces, electrically heated hot water appliances or other electrically heated appliances only.

5.28.12 Highly flammable products should be carried in suitably sealed containers.

5.28.13 Where bulkheads or decks required by 5.28.2, 5.28.3, 5.28.4, 5.28.5, 5.28.6, 5.28.8, 5.28.10 or 5.28.11 to be of "A", "B" or "F" class divisions are penetrated for the passage of electrical cables, pipes, trunks, ducts, etc., arrangements should be made to ensure that the fire integrity of the division is not impaired.

5.28.14 Air spaces enclosed behind ceilings, panellings or linings in accommodation spaces, service spaces and control stations should be divided by close-fitting draught stops spaced not more than 7 m apart.

5.28.15 Windows and skylights to machinery spaces should be as follows:

.1 where skylights can be opened, they should be capable of being closed from outside the space. Skylights containing glass panels should be fitted with external shutters of steel or other equivalent material which is permanently attached;

.2 glass or similar materials should not be fitted in machinery space boundaries. This does not preclude the use of wire-reinforced glass for skylights and glass in control rooms within the machinery spaces; and

.3 in skylights referred to in 5.28.15.1, wire-reinforced glass should be used.

5.28.16 Insulating materials in accommodation spaces, service spaces except domestic refrigerating compartments, control stations and machinery space should be non-combustible. The surface of insulation fitted on the internal boundaries of machinery spaces of category A should be impervious to oil or oil vapours.

5.28.17 Within compartments used for stowage of fish, combustible insulation should be protected by close-fitting cladding.

5.28.18 Notwithstanding the requirements of this section, the Competent Authority can accept "A-0" class divisions in lieu of "B-15" or "F" class divisions, having regard to the amount of combustible materials used in adjacent spaces.

5.29 Ventilation systems

5.29.1 Except as provided for in 5.30.2, means should be provided to stop fans and close main openings to ventilation systems from outside the spaces served.

5.29.2 Means should be provided for closing, from a safe position, the annular spaces around funnels.

5.29.3 Ventilation openings can be permitted in and under the doors in corridor bulkheads except that such openings should not be permitted in and under stairway enclosure doors. The openings should be provided only in the lower half of a door. Where such opening is in or under a door, the total net area of any such opening or openings should not exceed

0.05 m². When such opening is cut in a door, it should be fitted with a grille made of non-combustible material.

5.29.4 Ventilation ducts for machinery spaces of category A or galleys should not, in general, pass through accommodation spaces, service spaces or control stations. Where the Competent Authority permits this arrangement, the ducts should be constructed of steel or equivalent material and arranged to preserve the integrity of the divisions.

5.29.5 Ventilation ducts of accommodation spaces, service spaces or control stations should not, in general, pass through machinery spaces of category A or through galleys. Where the Competent Authority permits this arrangement, the ducts should be constructed of steel or equivalent material and arranged to preserve the integrity of the divisions.

5.29.6 Store-rooms containing appreciable quantities of highly flammable products should be provided with ventilation arrangements which are separate from other ventilation systems. Ventilation should be arranged at high and low levels and the inlets and outlets of ventilators should be positioned in safe areas. Suitable wire mesh guards to arrest sparks should be fitted over inlet and outlet ventilation openings.

5.29.7 Ventilation systems serving machinery spaces should be independent of systems serving other spaces.

5.29.8 Where trunks or ducts serve spaces on both sides of "A" class bulkheads or decks, dampers should be fitted so as to prevent the spread of fire and smoke between compartments. Manual dampers should be operable from both sides of the bulkhead or the deck. Where the trunks or ducts with a free cross-sectional area exceeding 0.02 m² pass through "A" class bulkheads or decks, automatic self-closing dampers should be fitted. Trunks serving compartments situated only on one side of such bulkheads should comply with 5.9.2.

5.30 Heating installations

5.30.1 Electric radiators should be fixed in position and so constructed as to reduce fire risks to a minimum. No such radiator should be fitted with an element so exposed that clothing, curtains or other similar materials can be scorched or set on fire by heat from the element.

5.30.2 Heating by means of open fires should not be permitted. Heating stoves and other similar appliances should be firmly secured and adequate protection and insulation against fire should be provided beneath and around such appliances and in way of their uptakes. Uptakes of stoves which burn solid fuel should be so arranged and designed as to minimize the possibility of becoming blocked by combustion products and should have a ready means for cleaning. Dampers for limiting draughts in uptakes should, when in the closed position, still leave an adequate area open. Spaces in which stoves are installed should be provided with ventilators of sufficient area to provide adequate combustion-air for the stove. Such ventilators should have no means of closure and their position should be such that no closing appliances in accordance with 2.9 are required.

5.30.3 Open-flame gas appliances, except cooking stoves and water heaters, should not be permitted. Spaces containing any such stoves or water heaters should have adequate ventilation to remove fumes and possible gas leakage to a safe place. All pipes conveying gas from container to stove or water heater should be of steel or other approved material. Automatic safety gas shut-off devices should be fitted to operate on loss of pressure in the gas main pipe or flame failure on any appliance.

5.31 Miscellaneous items*

5.31.1 Exposed surfaces within accommodation spaces, service spaces, control stations, corridor and stairway enclosures and the concealed surfaces behind bulkheads, ceilings, panellings and linings in accommodation spaces, service spaces, and control stations should have low flame-spread characteristics.†

5.31.2 All exposed surfaces of glass-reinforced plastic construction within accommodation and service spaces, control stations, machinery spaces of category A and other machinery spaces of similar fire risk should have the final lay-up layer of approved resin having inherent fire-retardant properties or be coated with an approved fire-retardant paint or be protected by non-combustible materials.

5.31.3 Paints, varnishes and other finishes used on exposed interior surfaces should not be capable of producing excessive quantities of smoke or toxic gases or vapours. The Competent Authority should be satisfied that they are not of a nature to offer undue fire hazard.

5.31.4 Primary deck coverings within accommodation and service spaces and control stations should be of approved material which will not readily ignite or give rise to toxic or explosive hazards at elevated temperatures.‡

5.31.5 In accommodation and service spaces and control stations, pipes penetrating "A" or "B" class divisions should be of approved materials having regard to the temperature that such divisions are required to withstand. Where the Competent Authority permits the conveying of oil

---

* Refer to the Guidance concerning the use of certain plastic materials, contained in recommendation 7 of attachment 3 to the Final Act of the 1993 Torremolinos Conference.

† Refer to Guidelines on the evaluation of fire hazard properties of materials, adopted by the Organization by resolution A.166(ES.IV) and the Recommendation on improved fire test procedures for surface flammability of bulkhead, ceiling and deck finish materials, adopted by the Organization by resolution A.653(16).

‡ For vessels, the deck of which are constructed of steel, refer to the Recommendation on fire test procedures for ignitability of primary deck coverings, adopted by the Organization by resolution A.687(17).

and combustible liquids through accommodation and service spaces, the pipes conveying oil or combustible liquids should be of an approved material having regard to the fire risk.

5.31.6 Materials readily rendered ineffective by heat should not be used for overboard scuppers, sanitary discharges and other outlets which are close to the waterline and where the failure of the material in the event of fire would give rise to danger of flooding.

5.31.7 All waste receptacles other than those used in fish processing should be constructed of non-combustible materials with no openings in the sides and bottom.

5.31.8 Machinery driving fuel oil transfer pumps, fuel oil unit pumps and other similar fuel pumps should be fitted with remote controls situated outside the space concerned, so that they can be stopped in the event of a fire arising in the space in which they are located.

5.31.9 Drip trays should be fitted, where necessary, to prevent oil leaking into bilges.

5.32 Storage of gas cylinders and dangerous materials

5.32.1 Cylinders for compressed, liquefied or dissolved gases should be clearly marked by means of prescribed identifying colours, should have a clearly legible identification of the name and chemical formula of their contents and should be properly secured.

5.32.2 Cylinders containing flammable or other dangerous gases and expended cylinders should be stored, properly secured, on open decks; and all valves, pressure regulators and pipes leading from such cylinders should be protected against damage. Cylinders should be protected against excessive variations in temperature, direct rays of the sun and accumulation of snow. However, the Competent Authority can permit such cylinders to be stored in compartments complying with the requirements of 5.32.3 to 5.32.5.

5.32.3 Spaces containing highly flammable liquids, such as volatile paints, paraffin, benzole, etc. and, where permitted, liquefied gas, should have direct access from open decks only. Pressure-adjusting devices and relief valves should exhaust within the compartment. Where boundary bulkheads of such compartments adjoin other enclosed spaces, they should be gas-tight.

5.32.4 Except as necessary for service within the space, electrical wiring and fittings should not be permitted within compartments used for the storage of highly flammable liquids or liquefied gases. Where such electrical fittings are installed, they should be to the satisfaction of the Competent Authority for use in a flammable atmosphere. Sources of heat should be kept clear of such spaces and "No Smoking" and "No Naked Light" notices should be displayed in a prominent position.

5.32.5 Separate storage should be provided for each type of compressed gas. Compartments used for the storage of such gases should not be used for storage of other combustible products nor for tools or objects not part of the gas distribution system. However, the Competent Authority may relax these requirements, considering the characteristics, volume and intended use of such compressed gases.

5.33 Means of escape

5.33.1 Stairways and ladders leading to and from all accommodation spaces and in spaces in which the crew is normally employed, other than machinery spaces, should be so arranged as to provide ready means of escape to the open deck and thence to the survival craft. In particular, in relation to these spaces:

.1 at all levels of accommodation, at least two widely separated means of escape should be provided which can include the normal means of access from each restricted space or group of spaces;

.2.1 below the weather deck, the means of escape should be a stairway and the second escape can be a trunk or a stairway; and

.2.2 above the weather deck, the means of escape should be stairways or doors to an open deck or a combination thereof. Where it is not practicable to fit stairways or doors, one of these means of escape can be by means of adequately sized portholes or hatches, protected, where necessary, against ice accretion;

.3 exceptionally, the Competent Authority can permit only one means of escape, due regard being paid to the nature and location of spaces and to the number of persons who normally might be accommodated or employed there;

.4 a corridor or part of a corridor from which there is only one route of escape should preferably not exceed 2.5 m in length and, in no case, be greater than 5 m in length; and

.5 the width and continuity of the means of escape should be to the satisfaction of the Competent Authority.

5.33.2 Two means of escape should be provided from every machinery space of category A, which should be as widely separated as possible. Vertical escapes should be by means of steel ladders. Where the size of the machinery spaces makes it impracticable, one of these means of escape can be omitted. In such cases, special consideration should be given to the remaining exit.

5.33.3 Lifts should not be considered as forming one of the required means of escape.

5.34 Automatic fire alarm and fire detection systems

Where the Competent Authority has permitted under 5.28.1 a combustible construction, or where otherwise appreciable amounts of combustible materials are used in the construction of accommodation spaces, service spaces and control stations, special consideration should be given to the installation of an automatic fire alarm and fire detection system in those spaces, having due regard to the size of those spaces, their arrangement and location relative to control stations as well as, where applicable, the flame-spread characteristics of the installed furniture.

5.35 Fire pumps

5.35.1 The minimum number and type of fire pumps to be fitted should be as follows:

   .1 one power pump not dependent upon the main machinery for its motive power; or

   .2 one power pump driven by main machinery provided that the propeller shafting can be readily disconnected or provided that a controllable-pitch propeller is fitted.

5.35.2 Sanitary, bilge, ballast, general service or any other pumps can be used as fire pumps if they comply with the requirements of this chapter and do not affect the ability to cope with pumping of the bilges. Fire pumps should be so connected that they cannot be used for pumping oil or other flammable liquids.

5.35.3 Centrifugal pumps or other pumps connected to the fire main through which backflow could occur should be fitted with non-return valves.

5.35.4 Vessels not fitted with a power-operated emergency fire pump and without a fixed fire-extinguishing system in the machinery spaces should be provided with additional fire-extinguishing means to the satisfaction of the Competent Authority.

5.35.5 When fitted, emergency power-operated fire pumps should be independently driven self-contained pumps either with their own prime mover and fuel supply fitted in an accessible position outside the compartment which contains the main fire pumps, or driven by a self-contained generator which can be an emergency generator of sufficient capacity and which is positioned in a safe place outside the engine-room and preferably above the working deck.

5.35.6 For any emergency fire pump, where fitted, the pump, sea-suction valves and other necessary valves should be operable from outside compartments containing main fire pumps in a position not likely to be cut off by a fire in those compartments.

Chapter V: part C

5.35.7 The total capacity (Q) of main power-operated fire pumps should be at least:

$$Q = (0.15 \sqrt{L(B+D)} + 2.25)^2 \text{ m}^3/\text{h}$$

where L, B and D are in metres.

5.35.8 Where two independent power-operated fire pumps are fitted, the capacity of each pump should not be less than 40% of the quantity required by 5.35.7.

5.35.9 When main power fire pumps are delivering the quantity of water required by 5.35.7 through the fire main, fire hoses and nozzles, the pressure maintained at any hydrant should be not less than 0.25 N/mm^2.

5.35.10 Where power-operated emergency fire pumps are delivering the maximum quantity of water through the jet required by 5.37.1, the pressure maintained at any hydrant should be to the satisfaction of the Competent Authority.

5.36 Fire mains

5.36.1 Where more than one hydrant is required to provide the number of jets required by 5.37.1, a fire main should be provided.

5.36.2 Materials readily rendered ineffective by heat should not be used for fire mains, unless adequately protected.

5.36.3 Where fire pump delivery pressure can exceed the designed working pressure of fire mains, relief valves should be fitted.

5.36.4 Fire mains should have no connections other than those required for fire fighting, except for the purpose of washing the deck and anchor chains or operation of bilge ejectors, subject to the efficiency of the fire-fighting system being maintained.

5.36.5 Where fire mains are not self-draining, suitable drain cocks should be fitted where frost damage may be expected.*

5.37 Fire hydrants, fire hoses and nozzles

5.37.1 Fire hydrants should be positioned so as to allow easy and quick connection of fire hoses and so that at least one jet can be directed into any part of the vessel which is normally accessible during navigation.

5.37.2 The jet required in 5.37.1 should be from a single length of fire hose.

5.37.3 In addition to the requirements of 5.37.1, machinery spaces of category A should be provided with at least one fire hydrant complete with

---

* Refer to the Guidance for precautions against freezing of fire mains, contained in recommendation 6 of attachment 3 to the Final Act of the 1993 Torremolinos Conference.

fire hose and dual-purpose nozzle. This fire hydrant should be located outside the space and near the entrance.

5.37.4 For every required fire hydrant, there should be one fire hose. At least one spare fire hose should be provided in addition to this requirement.

5.37.5 Single lengths of fire hose should not exceed 20 m.

5.37.6 Fire hoses should be of an approved material. Each fire hose should be provided with couplings and a dual-purpose nozzle.

5.37.7 Except where fire hoses are permanently attached to the fire main, the couplings of fire hoses and nozzles should be completely interchangeable.

5.37.8 The nozzles as required by 5.37.6 should be appropriate to the delivery capacity of the fire pumps fitted, but in any case should have a diameter of not less than 12 mm.

5.38 Fire extinguishers*

5.38.1 Fire extinguishers should be of approved types. The capacity of required portable fluid extinguishers should be not more than 13.5 l and not less than 9 l. Other extinguishers should not be in excess of the equivalent portability of the 13.5 l fluid extinguisher and should not be less than the fire-extinguishing equivalent of a 9 l fluid extinguisher. The Competent Authority should determine the equivalents of fire extinguishers.

5.38.2 Spare charges should be provided to the satisfaction of the Competent Authority.

5.38.3 Fire extinguishers containing an extinguishing medium which, in the opinion of the Competent Authority, either by itself or under expected conditions of use, gives off toxic gases in such quantities as to endanger persons should not be permitted.

5.38.4 Fire extinguishers should be periodically examined and subjected to such tests as the Competent Authority may require.

5.38.5 Normally, one of the portable fire extinguishers intended for use in any space should be stowed near an entrance to that space.

5.39 Portable fire extinguishers in control stations and accommodation and service spaces

5.39.1 A sufficient number of approved portable fire extinguishers should be provided in control stations and accommodation and service spaces to ensure that at least one extinguisher of a suitable type is readily available

---

* Refer to the Improved Guidelines for marine portable fire extinguishers, adopted by the Organization by resolution A.951(23).

Chapter V: part C

for use in any part of such spaces. The total number of extinguishers in these spaces, however, should not be less than three.

5.39.2 Spare charges should be provided to the satisfaction of the Competent Authority.

5.40 Fire-extinguishing appliances in machinery spaces

5.40.1 Spaces containing oil-fired boilers, fuel oil units or internal-combustion machinery having a total power output of not less than 750 kW should be provided with one of the following fixed fire-extinguishing systems, to the satisfaction of the Competent Authority:

.1 a pressure water-spraying installation;

.2 a fire-smothering gas installation;

.3 a fire-extinguishing installation using vapours from low-toxicity vaporizing liquids; or

.4 a fire-extinguishing installation using high-expansion foam.

5.40.2 New installations of halogenated hydrocarbon systems used as fire-extinguishing media should be prohibited on new and existing vessels.

5.40.3 Where the engine and boiler rooms are not entirely separated from each other or if fuel oil can drain from the boiler room into the engine-room, the combined engine and boiler rooms should be considered as one compartment.

5.40.4 Installations listed in 5.40.1 should be controlled from readily accessible positions outside such spaces not likely to be cut off by a fire in the protected space. Arrangements should be made to ensure the supply of power and water necessary for the operation of the system in the event of fire in the protected space.

5.40.5 Vessels which are constructed mainly or wholly of wood or fibre-reinforced plastic and fitted with oil-fired boilers or internal-combustion machinery which are decked in way of the machinery space with such material should be provided with one of the extinguishing systems referred to in 5.40.1.

5.40.6 In all machinery spaces of category A at least two portable extinguishers should be provided, of a type suitable for extinguishing fires involving fuel oil. Where such spaces contain machinery which has a total power output of not less than 250 kW, at least three such extinguishers should be provided. One of the extinguishers should be stowed near the entrance to the space.

5.40.7 Vessels having machinery spaces not protected by a fixed fire-extinguishing system should be provided with at least a 45 l foam extinguisher or its equivalent, suitable for fighting oil fires. Where the size of the machinery spaces makes this provision impracticable, the Competent Authority can accept an additional number of portable fire extinguishers.

5.41 Firefighters' outfits

The number of firefighters' outfits and their locations should be to the satisfaction of the Competent Authority.

5.42 Fire control plan

There should be a permanently exhibited fire control plan to the satisfaction of the Competent Authority.*

5.43 Ready availability of fire-extinguishing appliances

Fire-extinguishing appliances should be kept in good order and available for immediate use at all times.

5.44 Acceptance of substitutes

Where in this part any special type of appliance, apparatus, extinguishing medium or arrangement is specified, any other type of appliance, etc. can be allowed provided the Competent Authority is satisfied that it is not less effective.

Part D
Fire safety measures in vessels of 24 m in length and over but less than 45 m

5.45 Structural fire protection

5.45.1 The hull, superstructure, structural bulkheads, decks and deckhouses should be constructed of non-combustible materials. The Competent Authority may permit combustible construction provided the requirements of this section and the additional fire-extinguishing requirements of 5.40.5 are complied with.

---

* Refer to the Graphical symbols for shipboard fire control plans, adopted by the Organization by resolution A.952(23).

5.45.2 In vessels the hull of which is constructed of non-combustible materials, the decks and bulkheads separating machinery spaces of category A from accommodation spaces, service spaces or control stations should be constructed to "A-60" class standard where the machinery space of category A is not provided with a fixed fire-extinguishing system and to "A-0" class standard where such a system is fitted. Decks and bulkheads separating other machinery spaces from accommodation, service spaces and control stations should be constructed to "A-0" class standard. Decks and bulkheads separating control stations from accommodation and service spaces should be constructed to "B" class standard, insulated to the satisfaction of the Competent Authority.

5.45.3 In vessels the hull of which is constructed of combustible materials, the decks and bulkheads separating machinery spaces from accommodation spaces, service spaces or control stations should be constructed to "F" class or "B-15" class standard. In addition, machinery space boundaries should, as far as practicable, prevent the passage of smoke. Decks and bulkheads separating control stations from accommodation and service spaces should be constructed to "F" class standard.

5.45.4 In vessels the hull of which is constructed of non-combustible materials, such bulkheads should be of "B-0" class divisions.

5.45.5 In vessels the hull of which is constructed of combustible materials, such bulkheads should be of fire-retardant material to the satisfaction of the Competent Authority.

5.45.6 Any bulkhead of corridors serving accommodation spaces, service spaces and control stations should extend from deck to deck unless a continuous ceiling of the same class as the bulkhead is fitted on both sides of the bulkhead, in which case the bulkhead can terminate at the continuous ceiling.

5.45.7 Interior stairways serving accommodation spaces, service spaces or control stations should be of steel or other equivalent material. Such stairways connecting more than two decks should be within enclosures constructed of "F" class divisions in vessels the hull of which is constructed of combustible materials, or "B-15" class divisions in vessels the hull of which is constructed of non-combustible materials.

5.45.8 Doors and other closures of openings in bulkheads and decks referred to in 5.45.2 to 5.45.6, doors fitted to stairway enclosures referred to in 5.45.7 and doors fitted in engine and boiler casings, should be, as far as practicable, equivalent in resisting fire to the divisions in which they are fitted. Doors to machinery spaces of category A should be self-closing.

5.45.9 Lift trunks which pass through the accommodation and service spaces should be constructed of steel or equivalent material and should be provided with means of closing which will permit control of draught and smoke.

5.45.10 In vessels the hull of which is constructed of combustible materials, the boundary bulkheads and decks of spaces containing any emergency source of power and bulkheads and decks between galleys, paint rooms, lamp rooms or any store-rooms which contain appreciable quantities of highly flammable materials, and accommodation spaces, service spaces or control stations should be constructed of "F" class or "B-15" class divisions.

5.45.11 In vessels the hull of which is constructed of non-combustible materials, the decks and bulkheads referred to in 5.45.10 should be of "A" class divisions insulated to the satisfaction of the Competent Authority, having in mind the risk of fire, except that the Competent Authority can accept "B-15" class divisions between galley and accommodation spaces, service spaces and control stations when the galley contains electrically heated furnaces, electrically heated hot water appliances or other electrically heated appliances only.

5.45.12 Highly flammable products should be carried in suitably sealed containers.

5.45.13 Where bulkheads or decks required by 5.45.2, 5.45.3, 5.45.4, 5.45.5, 5.45.6, 5.45.8, 5.45.10 or 5.45.11 to be of "A", "B" or "F" class divisions are penetrated for the passage of electrical cables, pipes, trunks, ducts, etc., arrangements should be made to ensure that the fire integrity of the division is not impaired.

5.45.14 Air spaces enclosed behind ceilings, panellings or linings in accommodation spaces, service spaces and control stations should be divided by close-fitting draught stops spaced not more than 7 m apart.

5.45.15 Windows and skylights to machinery spaces should be as follows:

.1 where skylights can be opened, they should be capable of being closed from outside the space. Skylights containing glass panels should be fitted with external shutters of steel or other equivalent material which is permanently attached;

.2 glass or similar materials should not be fitted in machinery space boundaries. This does not preclude the use of wire-reinforced glass for skylights and glass in control rooms within the machinery spaces; and

.3 in skylights referred to in 5.45.15.1, wire-reinforced glass should be used.

5.45.16 Insulating materials in accommodation spaces, service spaces except domestic refrigerating compartments, control stations and machinery space should be non combustible. The surface of insulation fitted on the internal boundaries of machinery spaces of category A should be impervious to oil or oil vapours.

5.45.17 Within compartments used for stowage of fish, combustible insulation should be protected by close-fitting cladding.

5.45.18 Notwithstanding the requirements of this section, the Competent Authority can accept "A-0" class divisions in lieu of "B-15" or "F" class divisions, having regard to the amount of combustible materials used in adjacent spaces.

5.46 Ventilation systems

5.46.1 Except as provided for in 5.47.2, means should be provided to stop fans and close main openings to ventilation systems from outside the spaces served.

5.46.2 Means should be provided for closing, from a safe position, the annular spaces around funnels.

5.46.3 Ventilation openings can be permitted in and under the doors in corridor bulkheads except that such openings should not be permitted in and under stairway enclosure doors. The openings should be provided only in the lower half of a door. Where such opening is in or under a door, the total net area of any such opening or openings should not exceed 0.05 m^2. When such opening is cut in a door, it should be fitted with a grille made of non-combustible material.

5.46.4 Ventilation ducts for machinery spaces of category A or galleys should not, in general, pass through accommodation spaces, service spaces or control stations. Where the Competent Authority permits this arrangement, the ducts should be constructed of steel or equivalent material and arranged to preserve the integrity of the divisions.

5.46.5 Ventilation ducts of accommodation spaces, service spaces or control stations should not, in general, pass through machinery spaces of category A or through galleys. Where the Competent Authority permits this arrangement, the ducts should be constructed of steel or equivalent material and arranged to preserve the integrity of the divisions.

5.46.6 Store-rooms containing appreciable quantities of highly flammable products should be provided with ventilation arrangements which are separate from other ventilation systems. Ventilation should be arranged at high and low levels and the inlets and outlets of ventilators should be positioned in safe areas. Suitable wire mesh guards to arrest sparks should be fitted over inlet and outlet ventilation openings.

5.46.7 Ventilation systems serving machinery spaces should be independent of systems serving other spaces.

5.46.8 Where trunks or ducts serve spaces on both sides of "A" class bulkheads or decks, dampers should be fitted so as to prevent the spread of fire and smoke between compartments. Manual dampers should be operable from both sides of the bulkhead or the deck. Where the trunks or ducts with a free cross-sectional area exceeding 0.02 m^2 pass through "A" class bulkheads or decks, automatic self-closing dampers should be fitted. Trunks serving compartments situated only on one side of such bulkheads should comply with 5.9.2.

5.47 Heating installations

5.47.1 Electric radiators should be fixed in position and so constructed as to reduce fire risks to a minimum. No such radiator should be fitted with an element so exposed that clothing, curtains or other similar materials can be scorched or set on fire by heat from the element.

5.47.2 Heating by means of open fires should not be permitted. Heating stoves and other similar appliances should be firmly secured and adequate protection and insulation against fire should be provided beneath and around such appliances and in way of their uptakes. Uptakes of stoves which burn solid fuel should be so arranged and designed as to minimize the possibility of becoming blocked by combustion products and should have a ready means for cleaning. Dampers for limiting draughts in uptakes should, when in the closed position, still leave an adequate area open. Spaces in which stoves are installed should be provided with ventilators of sufficient area to provide adequate combustion-air for the stove. Such ventilators should have no means of closure and their position should be such that no closing appliances in accordance with 2.9 are required.

5.47.3 Open-flame gas appliances, except cooking stoves and water heaters, should not be permitted. Spaces containing any such stoves or water heaters should have adequate ventilation to remove fumes and possible gas leakage to a safe place. All pipes conveying gas from container to stove or water heater should be of steel or other approved material. Automatic safety gas shut-off devices should be fitted to operate on loss of pressure in the gas main pipe or flame failure on any appliance.

5.48 Miscellaneous items*

5.48.1 Exposed surfaces within accommodation spaces, service spaces, control stations, corridor and stairway enclosures and the concealed surfaces behind bulkheads, ceilings, panellings and linings in accommodation spaces, service spaces, and control stations should have low flame-spread characteristics, or to be of fire-retardant materials to the satisfaction of the Competent Authority.†

5.48.2 All exposed surfaces of glass-reinforced plastic construction within accommodation and service spaces, control stations, machinery spaces of category A and other machinery spaces of similar fire risk should have the final lay-up layer of approved resin having inherent fire-retardant proper-

---

* Refer to the Guidance concerning the use of certain plastic materials, contained in recommendation 7 of attachment 3 to the Final Act of the 1993 Torremolinos Conference.
† Refer to the Guidelines on the evaluation of fire properties of materials, adopted by the Organization by resolution A.166(ES.IV), and the Recommendation on improved fire test procedures for surface flammability of bulkhead, ceiling and deck finish materials, adopted by the Organization by resolution A.653(16).

Chapter V: part D

ties or be coated with an approved fire-retardant paint or be protected by non-combustible materials.

5.48.3 Paints, varnishes and other finishes used on exposed interior surfaces should not be capable of producing excessive quantities of smoke or toxic gases or vapours. The Competent Authority should be satisfied that they are not of a nature to offer undue fire hazard.

5.48.4 Primary deck coverings within accommodation and service spaces and control stations should be of approved material which will not readily ignite or give rise to toxic or explosive hazards at elevated temperatures.*

5.48.5 In accommodation and service spaces and control stations, pipes penetrating "A" or "B" class divisions should be of approved materials having regard to the temperature that such divisions are required to withstand. Where the Competent Authority permits the conveying of oil and combustible liquids through accommodation and service spaces, the pipes conveying oil or combustible liquids should be of an approved material having regard to the fire risk.

5.48.6 Materials readily rendered ineffective by heat should not be used for overboard scuppers, sanitary discharges and other outlets which are close to the waterline and where the failure of the material in the event of fire would give rise to danger of flooding.

5.48.7 All waste receptacles other than those used in fish processing should be constructed of non-combustible materials with no openings in the sides and bottom.

5.48.8 Machinery driving fuel oil transfer pumps, fuel oil unit pumps and other similar fuel pumps should be fitted with remote controls situated outside the space concerned, so that they can be stopped in the event of a fire arising in the space in which they are located.

5.48.9 Drip trays should be fitted, where necessary, to prevent oil leaking into bilges.

5.49 Storage of gas cylinders and dangerous materials

5.49.1 Cylinders for compressed, liquefied or dissolved gases should be clearly marked by means of prescribed identifying colours, should have a clearly legible identification of the name and chemical formula of their contents and should be properly secured.

5.49.2 Cylinders containing flammable or other dangerous gases and expended cylinders should be stored, properly secured, on open decks; and all valves, pressure regulators and pipes leading from such cylinders

---

* For vessels the decks of which are constructed of steel, refer to the Recommendation on fire test procedures for ignitability of primary deck coverings, adopted by the Organization by resolution A.687(17).

should be protected against damage. Cylinders should be protected against excessive variations in temperature, direct rays of the sun, and accumulation of snow. However, the Competent Authority can permit such cylinders to be stored in compartments complying with the requirements of 5.49.3 to 5.49.5.

5.49.3 Spaces containing highly flammable liquids, such as volatile paints, paraffin, benzole, etc. and, where permitted, liquefied gas, should have direct access from open decks only. Pressure-adjusting devices and relief valves should exhaust within the compartment. Where boundary bulkheads of such compartments adjoin other enclosed spaces, they should be gastight.

5.49.4 Except as necessary for service within the space, electrical wiring and fittings should not be permitted within compartments used for the storage of highly flammable liquids or liquefied gases. Where such electrical fittings are installed, they should be to the satisfaction of the Competent Authority for use in a flammable atmosphere. Sources of heat should be kept clear of such spaces and "No Smoking" and "No Naked Light" notices should be displayed in a prominent position.

5.49.5 Separate storage should be provided for each type of compressed gas. Compartments used for the storage of such gases should not be used for storage of other combustible products nor for tools or objects not part of the gas distribution system. However, the Competent Authority may relax these requirements considering the characteristics, volume and intended use of such compressed gases.

5.50 Means of escape

5.50.1 Stairways and ladders leading to and from all accommodation spaces and in spaces in which the crew is normally employed, other than machinery spaces, should be so arranged as to provide ready means of escape to the open deck and thence to the survival craft. In particular, in relation to these spaces:

- .1 at all levels of accommodation, at least two widely separated means of escape should be provided which can include the normal means of access from each restricted space or group of spaces;
- .2.1 below the weather deck, the means of escape should be a stairway and the second escape can be a trunk or a stairway; and
- .2.2 above the weather deck, the means of escape should be stairways or doors to an open deck or a combination thereof. Where it is not practicable to fit stairways or doors, one of these means of escape can be by means of adequately sized portholes or hatches, protected, where necessary, against ice accretion;

- .3 exceptionally, the Competent Authority can permit only one means of escape, due regard being paid to the nature and location of spaces and to the number of persons who normally might be accommodated or employed there;
- .4 a corridor or part of a corridor from which there is only one route of escape should preferably not exceed 2.5 m in length and, in no case, be greater than 5 m in length; and
- .5 the width and continuity of the means of escape should be to the satisfaction of the Competent Authority.

5.50.2 Two means of escape should be provided from every machinery space of category A, which should be as widely separated as possible. Vertical escapes should be by means of steel ladders. Where the size of the machinery spaces makes it impracticable, one of these means of escape can be omitted. In such cases, special consideration should be given to the remaining exit.

5.50.3 Lifts should not be considered as forming one of the required means of escape.

5.51 Automatic fire alarm and fire detection systems

Where the Competent Authority has permitted under 5.45.1 a combustible construction, or where otherwise appreciable amounts of combustible materials are used in the construction of accommodation spaces, service spaces and control stations, special consideration should be given to the installation of an automatic fire alarm and fire detection system in those spaces, having due regard to the size of those spaces, their arrangement and location relative to control stations as well as, where applicable, the flame-spread characteristics of the installed furniture.

5.52 Fire pumps

5.52.1 The minimum number and type of fire pumps to be fitted should be as follows:

- .1 one power pump not dependent upon the main machinery for its motive power; or
- .2 one power pump driven by main machinery provided that the propeller shafting can be readily disconnected or provided that a controllable-pitch propeller is fitted.

5.52.2 Sanitary, bilge, ballast, general service or any other pumps can be used as fire pumps if they comply with the requirements of this chapter and do not affect the ability to cope with pumping of the bilges. Fire pumps should be so connected that they cannot be used for pumping oil or other flammable liquids.

5.52.3 Centrifugal pumps or other pumps connected to the fire main through which backflow could occur should be fitted with non-return valves.

5.52.4 Vessels not fitted with a power-operated emergency fire pump and without a fixed fire-extinguishing system in the machinery spaces should be provided with additional fire-extinguishing means to the satisfaction of the Competent Authority.

5.52.5 When fitted, emergency power-operated fire pumps should be independently driven self-contained pumps either with their own prime mover and fuel supply fitted in an accessible position outside the compartment which contains the main fire pumps, or be driven by a self-contained generator which can be an emergency generator of sufficient capacity and which is positioned in a safe place outside the engine-room and preferably above the working deck.

5.52.6 For any emergency fire pump, where fitted, the pump, sea-suction valves and other necessary valves should be operable from outside compartments containing main fire pumps in a position not likely to be cut off by a fire in those compartments.

5.52.7 The total capacity (Q) of main power-operated fire pumps should be at least:

$$Q = (0.15 \sqrt{L(B+D)} + 2.25)^2 \text{ m}^3/\text{h}$$

where L, B and D are in metres.

5.52.8 Where two independent power-operated fire pumps are fitted, the capacity of each pump should not be less than 40% of the quantity required by 5.52.7.

5.52.9 When main power fire pumps are delivering the quantity of water required by 5.52.7 through the fire main, fire hoses and nozzles, the pressure maintained at any hydrant should be not less than 0.25 N/mm².

5.52.10 Where power-operated emergency fire pumps are delivering the maximum quantity of water through the jet required by 5.54.1, the pressure maintained at any hydrant should be to the satisfaction of the Competent Authority.

5.53 Fire mains

5.53.1 Where more than one hydrant is required to provide the number of jets required by 5.54.1, a fire main should be provided.

5.53.2 Materials readily rendered ineffective by heat should not be used for fire mains, unless adequately protected.

5.53.3 Where fire pump delivery pressure can exceed the designed working pressure of fire mains, relief valves should be fitted.

5.53.4 Fire mains should have no connections other than those required for fire fighting, except for the purpose of washing the deck and anchor chains or operation of bilge ejectors, subject to the efficiency of the fire-fighting system being maintained.

5.53.5 Where fire mains are not self-draining, suitable drain cocks should be fitted where frost damage may be expected.*

5.54 Fire hydrants, fire hoses and nozzles

5.54.1 Fire hydrants should be positioned so as to allow easy and quick connection of fire hoses and so that at least one jet can be directed into any part of the vessel which is normally accessible during navigation.

5.54.2 The jet required in 5.54.1 should be from a single length of fire hose.

5.54.3 In addition to the requirements of 5.54.1, machinery spaces of category A should be provided with at least one fire hydrant complete with fire hose and dual-purpose nozzle. This fire hydrant should be located outside the space and near the entrance.

5.54.4 For every required fire hydrant, there should be one fire hose. At least one spare fire hose should be provided in addition to this requirement.

5.54.5 Single lengths of fire hose should not exceed 20 m.

5.54.6 Fire hoses should be of an approved material. Each fire hose should be provided with couplings and a dual-purpose nozzle.

5.54.7 Except where fire hoses are permanently attached to the fire main, the couplings of fire hoses and nozzles should be completely interchangeable.

5.54.8 The nozzles as required by 5.54.6 should be appropriate to the delivery capacity of the fire pumps fitted, but in any case should have a diameter of not less than 12 mm.

5.55 Fire extinguishers[†]

5.55.1 Fire extinguishers should be of approved types. The capacity of required portable fluid extinguishers should be not more than 13.5 *l* and not less than 9 *l*. Other extinguishers should not be in excess of the equivalent portability of the 13.5 *l* fluid extinguisher and should not be less than the fire-extinguishing equivalent of a 9 *l* fluid extinguisher. The Competent Authority should determine the equivalents of fire extinguishers.

---

* Refer to the Guidance for precautions against freezing of fire mains, contained in recommendation 6 of attachment 3 to the Final Act of the 1993 Torremolinos Conference.
† Refer to the Improved Guidelines for marine portable fire extinguishers, adopted by the Organization by resolution A.951(23).

Fire protection, detection, extinction and fighting

5.55.2 Spare charges should be provided to the satisfaction of the Competent Authority.

5.55.3 Fire extinguishers containing an extinguishing medium which, in the opinion of the Competent Authority, either by itself or under expected conditions of use, gives off toxic gases in such quantities as to endanger persons should not be permitted.

5.55.4 Fire extinguishers should be periodically examined and subjected to such tests as the Competent Authority may require.

5.55.5 Normally, one of the portable fire extinguishers intended for use in any space should be stowed near an entrance to that space.

5.56 Portable fire extinguishers in control stations and accommodation and service spaces

5.56.1 A sufficient number of approved portable fire extinguishers should be provided in control stations and accommodation and service spaces to ensure that at least one extinguisher of a suitable type is readily available for use in any part of such spaces. The total number of extinguishers in these spaces, however, should not be less than three.

5.56.2 Spare charges should be provided to the satisfaction of the Competent Authority.

5.57 Fire-extinguishing appliances in machinery spaces

5.57.1 Spaces containing oil-fired boilers, fuel oil units or internal-combustion machinery having a total power output of not less than 750 kW should be provided with one of the following fixed fire-extinguishing systems, to the satisfaction of the Competent Authority:

    .1   a pressure water-spraying installation;

    .2   a fire-smothering gas installation;

    .3   a fire-extinguishing installation using vapours from low-toxicity vaporizing liquids; or

    .4   a fire-extinguishing installation using high-expansion foam.

5.57.2 New installations of halogenated hydrocarbon systems used as fire-extinguishing media should be prohibited on new and existing vessels.

5.57.3 Where the engine and boiler rooms are not entirely separated from each other or if fuel oil can drain from the boiler room into the engine-room, the combined engine and boiler rooms should be considered as one compartment.

5.57.4 Installations listed in 5.57.1 should be controlled from readily accessible positions outside such spaces not likely to be cut off by a fire in the protected space. Arrangements should be made to ensure the supply of power and water necessary for the operation of the system in the event of fire in the protected space.

5.57.5 Vessels which are constructed mainly or wholly of wood or fibre-reinforced plastic and fitted with oil-fired boilers or internal-combustion machinery which are decked in way of the machinery space with such material should be provided with one of the extinguishing systems referred to in 5.57.1.

5.57.6 In all machinery spaces of category A at least two portable extinguishers should be provided, of a type suitable for extinguishing fires involving fuel oil. Where such spaces contain machinery which has a total power output of not less than 250 kW, at least three such extinguishers should be provided. One of the extinguishers should be stowed near the entrance to the space.

5.57.7 Vessels having machinery spaces not protected by a fixed fire-extinguishing system should be provided with at least a 45 l foam extinguisher or its equivalent, suitable for fighting oil fires. Where the size of the machinery spaces makes this provision impracticable, the Competent Authority can accept an additional number of portable fire extinguishers.

5.58 Firefighters' outfits

The number of firefighters' outfits and their locations should be to the satisfaction of the Competent Authority.

5.59 Fire control plan

The Competent Authority may dispense with this requirement.

5.60 Ready availability of fire-extinguishing appliances

Fire-extinguishing appliances should be kept in good order and available for immediate use at all times.

5.61 Acceptance of substitutes

Where in this part any special type of appliance, apparatus, extinguishing medium or arrangement is specified, any other type of appliance, etc. can be allowed provided the Competent Authority is satisfied that it is not less effective.

Chapter VI
Protection of the crew

6.1 General protective measures

6.1.1 An adequate number of lifelines, safety belts, bosun's chairs and stages should be provided.

6.1.2 A lifeline system should be designed to be effective for all needs and the necessary wires, ropes, shackles, eye bolts and cleats should be provided.

6.1.3 Where there is a danger of personnel falling through deck openings, the coamings or sills should have a suitable height; otherwise, such openings should be provided with suitable guards, such as hinged or portable railings or netting. The Competent Authority should take into consideration the position and operational use of small openings such as fish and ice scuttles before deciding whether or not they should be provided with guards.

6.1.4 Where there is a danger of personnel falling through skylights or other similar openings, such openings should be fitted with protective bars not more than 350 mm apart.

6.1.5 The surface of decks throughout a vessel should be specially designed or treated to minimize the possibility of personnel slipping. In particular, the decks and working spaces on board, such as machinery spaces, galleys and fish-handling and winch areas as well as deck areas at the foot and head of ladders and just outside the doors, should be specially prepared or designed as anti-skid surfaces.

6.1.6 The Competent Authority should be satisfied that, consistent with safety and operational procedures, working arrangements on board should provide for adequate rest periods for the crew.

6.2 Deck openings

6.2.1 Hinged covers of hatchways, manholes and other openings should be protected against accidental closing. In particular, heavy covers on escape hatches should be equipped with counterweights. The hatch should be so constructed that it can be opened from each side of the cover.

Chapter VI

6.2.2 Dimensions of access hatches should not be less than 600 mm by 600 mm or 600 mm in diameter.

6.2.3 Where practical, handholds should be provided above the level of the deck over escape openings.

6.2.4 External hatches and doors should be closed when the vessel is at sea. All openings occasionally required to be kept open during fishing and which may lead to flooding should be closed immediately if such danger of filling occurs with subsequent loss of buoyancy and stability.

6.3 Bulwarks, rails and guards

6.3.1 Efficient bulwarks or guardrails should be fitted on all exposed parts of the working deck and on superstructure and deckhouse decks if they are working platforms. The height above deck of bulwarks or guardrails should be at least 1 m, provided that, where this would interfere with the normal operation of the vessel, a lesser height may be approved by the Competent Authority if adequate protection is provided.

6.3.2 The minimum vertical distance from the deepest operating waterline to the lowest point of the top of the bulwark, or to the edge of the working deck if guardrails are fitted, should ensure adequate protection for the crew from water shipped on deck, taking into account the sea states and the weather conditions in which the vessel may operate, the area of operation, type of vessel and its method of fishing.*

6.3.3 Clearance below the lowest course of guardrails should not exceed 230 mm. Other courses should not be more than 380 mm apart, and the distance between stanchions should not be more than 1.5 m. In a vessel with rounded gunwales, guardrail supports should be placed on the flat of the deck. Rails should be free from sharp points, edges and corners and should be of adequate strength.

6.3.4 Satisfactory means in the form of guardrails, lifelines, gangways or underdeck passages, etc. should be provided for the protection of the crew in getting to and from their quarters, machinery spaces and other working spaces. Storm rails should be fitted on the outside of all deckhouses and casings.

6.3.5 A stern trawler should be provided with doors, gates or other suitable protective arrangements at the top of the stern ramp at the same height as the adjacent bulwark or guardrails. When such protection is not in position, a chain or other means of protection should be provided across the ramp.

---

* Refer to the Guidance on a method of calculation of the minimum distance from the deepest operating waterline to the lowest point of the top of the bulwark or to the edge of the working deck, contained in recommendation 8 of attachment 3 to the Final Act of the 1993 Torremolinos Conference.

6.3.6 Where a section of a bulwark or guard rail has to be removed or swung open to facilitate the fishing operation, as in the case of longline fishing, protection for the crew should be provided at the opening. When it is not practical to reinstate the bulwark or guardrail between hauling operations, chains or other means of protection should be provided across the opening. Where chains are to be fitted, the upper chain should be level with the upper edge of the bulwark or guardrail and at least one other chain should be fitted midway between the upper chain and the deck.

6.4 Stairways and ladders

6.4.1 Stairways and ladders should be provided for safe working at sea and in port. They should be of adequate size and strength. Means of access to holds, 'tween-decks, bunkers and similar parts of a vessel should consist of fixed ladders or stairs.

6.4.2 Stairways of more than 1 m in height should have handrails on both sides.

6.4.3 Treads of stairways should be flat and specially prepared to minimize slipping.

6.4.4 Fixed vertical ladders should be so situated as to be protected from damage and should be so fitted as to provide clearance of 150 mm behind. The rungs of steel vertical ladders should be made of square section steel bars with the sharp edge upwards. Where ladders are constructed with stringers, the rungs should pass through the stringers. Handholds should be provided where rungs or stringers are not suitable for this purpose.

6.4.5 Emergency escape ladders should normally be fixed, but may be portable provided that they are stowed adjacent to the escape and that they can be secured in place when required without tools or mechanical aids.

6.4.6 Ladders in machinery and boiler spaces should preferably be at least 450 mm wide.

6.5 Accommodation ladders and gangways

6.5.1 An accommodation ladder, gangway or similar appliances should normally be provided to ensure sufficiently safe and convenient access to the vessel.

6.5.2 If an accommodation ladder or gangway is not practicable, a substantial straight ladder, of adequate length and extending at least 900 mm above the upper landing surface, should be provided. Where conditions are such that a ladder cannot be used, a pilot ladder meeting the provisions of annex VI to this part of the Code should be provided.

Chapter VI

6.5.3 Accommodation ladders and gangways should be of reliable material, good construction and adequate strength, and be securely installed.

6.5.4 Accommodation ladders and gangways should be at least 550 mm wide and be fitted with railings at least 1 m high measured perpendicularly to the appliance on both sides, consisting of two rails or taut ropes, wires or chains about 500 mm apart and supported by stanchions not more than 2 m apart which should be designed to be secured against inadvertent dislodgement.

6.5.5 Accommodation ladders should be provided with hooks or other suitable fastenings for adequate support and securing against displacement or slipping and be able to be adjusted to the height of the landing place.

6.5.6 When a fixed-tread accommodation ladder is fitted, cleated duckboards should be provided which can be secured over the front edges of ladder steps to form a gangway when the ladder angle is low enough to require this for safety.

6.5.7 Gangways should be fitted with cleats (transverse treads) at suitable stepping intervals and for the full width of the gangway.

6.5.8 Turntables of gangways which pivot or swivel horizontally on a platform should be adequately protected by railings or ropes.

6.5.9 The lower end of accommodation ladders or gangways should have suitable angle plates or guards to cover wheels or rollers and to serve as a runway to the landing surface.

6.5.10 Where the shipboard ends of the means of access rest on the top of the bulwark, steps equipped with at least one handrail of 900 mm in height should be provided which can be secured between the top of the bulwark and the vessel's deck.

6.6 Galleys

6.6.1 Adequate grabrails should be fitted.

6.6.2 Dangerous parts of food-processing machinery should be fitted with permanent safety guards.

6.6.3 Cooking stoves should be fitted with guards to retain cooking utensils.

6.6.4 Galley floor areas should be adequately drained. The anti-skid surface referred to in paragraph 6.1.5 should be designed to facilitate drainage.

6.6.5 Machinery, such as pumps and domestic refrigeration compressor units, situated in the galley should be fitted with permanent safety guards.

6.7 Deck machinery, tackle and lifting gear

General

6.7.1 All elements of a fishing gear system, including warping heads, winches, warps, wires, tackle, nets, etc., should be designed, arranged and installed to provide safe and convenient operation. Insofar as is possible, such components should be of a suitable strength so that, in the event of an overload strain, the failure will occur on the designated weak link in the system. All crew members should be made aware of the designated weak link in the system.

6.7.2 Warp guards should be fitted where practicable between warp lead rollers.

6.7.3 Sheaves and rollers should be guarded, where practicable.

6.7.4 Chains or other suitable devices should be provided for "stoppering off".

6.7.5 Wires and warps provided should be of adequate strength for the anticipated loads.

6.7.6 Where practicable, provision should be made to stop trawl boards swinging inboard, such as the fitting of a portable prevention bar at the gallows aperture or other equally effective means.

6.7.7 Lifting and running parts of the fishing gear should be of adequate strength for the anticipated loads.

6.7.8 Provision should be made for the stowage of bulky netting to allow for drainage and to prevent lateral movement. The stowage area should be of adequate dimensions to keep the centre of gravity of the stowed net to a minimum and to allow for the crew to work in safety when flaking down nets.

6.7.9 Moving parts of winches, line and net hauling equipment and of warp and chain leads which may present a hazard should be, as far as practicable, adequately guarded and fenced.

6.7.10 Controls of winches, line and net hauling equipment should be so placed that winch operators have ample room for their unimpeded operation and have as unobstructed a view as possible of the working area. Where possible, control handles should be arranged to return to the stop position when released and be provided, where necessary, with a suitable locking device in the stop/neutral position, to prevent accidental movements or displacement or unauthorized use. In general, winches and hauling equipment for fishing gear should be fitted with safety devices designed to prevent accidents.

6.7.11 The arrangement of the safety devices should also ensure that an emergency stop would be activated if a person is pulled towards a winch or other hauling equipment.

Chapter VI

6.7.12 Quick-release devices should preferably be fitted, in the case of beam trawling and in purse seining, that can be activated in an emergency from the wheelhouse and at the main control station if not in the wheelhouse.

6.7.13 The design and construction of winches, line and net hauling equipment should be such that the maximum effort necessary for operating handwheels, handles, crank handles, levers, etc. should not exceed 160 N and in the case of pedals not exceed 320 N.

Winches

6.7.14 The design of winch systems should ensure that, when power is supplied to the winch, the control valves and/or levers would always be in the stop/neutral position.

6.7.15 Winches should be provided with means to prevent overhoisting and to prevent the accidental release of a load, if power supply fails. Where practicable, winches with wire storage drums should be fitted to avoid the need to use warping heads.

6.7.16 Winches should be equipped with brakes capable of effectively arresting and holding the safe working load. Brakes should be proof tested before installation with a static load suitably in excess of the maximum safe working load to the satisfaction of the Competent Authority. Brakes should be provided with simple and easily accessible means of adjustment. Every winch drum which could be uncoupled from the drive should be furnished with a separate brake independent of the brake connected with the drive.

6.7.17 Where manually-operated "guiding on" gear is installed, the operating wheels should be without open spokes or protrusions that could cause injury to the operator and should be capable of being disengaged when the warps are paying out. Preferably the "guiding on" gear should be capable of being disengaged when the warps are paying out.

6.7.18 Where practicable, winches should be reversible.

6.7.19 Winch barrels should be provided with means for fastening wire ends, for instance clamps, shackles or other equally effective method, which should be so designed as to prevent kinking of the wires.

6.7.20 Where a fishing winch is provided with local and remote controls, these should be so arranged as to prevent simultaneous operation. The operator should have a clear view of the winch and adjacent area from either position. An emergency cut-off should be provided at the winch and at the remote station as well as in the wheelhouse.

6.7.21 Where a fishing winch is controlled from the wheelhouse, an emergency control switch at the winch should be provided. Where a second control at the winch is required by the Competent Authority, the arrangement should be such as to make simultaneous control from both control positions impossible, as well as to show which control position is in operation. Where necessary, emergency switches for winches should be provided remote from the winch to protect crew members working in places which are dangerous for operation of warps and trawl boards. Where a fishing winch is controlled from the bridge, the arrangements should be such that the operator has a direct or televised clear view of the winch and adjacent area.

Line and net hauling equipment

6.7.22 Line and net hauling equipment should be fitted with devices to ensure that the designated safe working load is not exceeded. Such devices should be tested to the satisfaction of the Competent Authority.

6.7.23 Where line and net hauling equipment is intended to be blocked or braked in the stop position, the arrangements should be tested to the satisfaction of the Competent Authority.

6.7.24 Where line and net hauling equipment is controlled from the wheelhouse or from a position remote from the equipment, means should be provided at the equipment to stop hauling and/or shooting in an emergency. In like manner, when the main controls are at the equipment, means should be provided in the wheelhouse to stop it in an emergency.

6.7.25 The arrangement of the safety devices should also ensure that an emergency stop would be activated if a person is pulled towards a line or net hauling equipment.

Lifting gear

6.7.26 Cranes should be well constructed of sound material and the design should conform with national standards that may be appropriate. The cranes should be tested to the satisfaction of the Competent Authority and the crane should be marked with the designated maximum safe working load. In the case of a crane fitted with an extendable jib, the safe working load at various radii should be clearly marked as close as practical to the operating controls.

6.7.27 In general, cranes adapted to carry net hauling equipment should be so designed that, in the fail-safe condition, the hanging point of the jib should not be too high or extend so far beyond the bulwark that retrieval of fishing gear or equipment would endanger the crew.

6.7.28 The braking or blocking arrangements of a crane should be tested to at least 1.5 times the designated safe working load to the satisfaction of the Competent Authority.

6.7.29 Lifting and hoisting appliances, as well as derricks and similar equipment including all parts of the working gear thereof, whether fixed or movable, and all plant should be of good construction, reliable material, adequate strength and free from patent defect. They should be adequately and suitably anchored, supported or suspended, having regard to the purpose for which they are used, and should be marked with the safe working load. They should have easy access for maintenance. Guards should be provided to prevent any undesirable movement of lifted or hoisted parts, such as cod end or fishing gear, which could present danger to the crew.

6.7.30 Lifting and hoisting appliances, as well as derricks, should be protected from overhoisting.

6.7.31 The Competent Authority should ensure that lifting and hoisting appliances, as well as derricks, should be tested at least every two years and the results entered in the record of the vessel.

6.7.32 No such appliance of a kind referred to in 6.7.27, nor any part or working gear thereof, should be taken into use for the first time or after it has undergone any substantial repair unless it has been tested and the result entered in the record of the vessel.

6.8 Lighting in working spaces and areas

6.8.1 All companionways, door or other normal access openings should be provided with artificial lighting on both sides of the opening to facilitate safe passage.

6.8.2 In all passageways and in those working areas not adequately lighted by natural lighting, artificial lighting should be provided to the satisfaction of the Competent Authority. Particular attention should be paid to rule 20(b) of the International Regulations for Preventing Collisions at Sea, 1972.

6.8.3 Glare, dazzle or sudden contrasts of illumination should be eliminated to the extent possible, taking into consideration the need for effective lighting for the safety of the crew working on deck.

6.8.4 Provision should be made for some form of emergency lighting which is independent of the normal supply.

6.8.5 Portable lights should be provided as necessary and fitted with heavy-duty cables, bulb guards and lanyards. Portable lights for use in spaces which may contain explosive gases should be either explosion-proof or otherwise intrinsically safe to the satisfaction of the Competent Authority.

6.8.6 Where necessary to prevent danger, electric lamps should be protected by guards.

6.8.7 In order to avoid the stroboscopic effect of fluorescent lighting, double-tube lamps should be used to illuminate working spaces with revolving machinery.

6.9 Ventilation in working and storage spaces

6.9.1 Enclosed working spaces, machinery spaces and spaces used for storage, in particular, of paints, oils, solvents and wet batteries should be adequately ventilated and in accordance with 5.9, 5.29 and 5.46.

6.9.2 Where necessary to safeguard personnel, workplaces and storage spaces should be provided with an adequate system of heating and/or cooling.

6.10 Dangerous areas

6.10.1 Dangerous spaces or entrances thereto should be properly illuminated and marked and have warning signs prominently posted. Retro-reflective and fluorescent materials may be used to increase the conspicuousness. A notice should also be posted if a first-aid procedure is appropriate.

6.10.2 A notice should be prominently displayed below radar and radio aerials warning of danger, with an instruction that the authority of a responsible officer should be obtained before work is done in the vicinity. A notice should also be prominently displayed at or near the operating controls of radar and radio equipment warning that, before starting up the equipment, it should be ascertained that no one is working aloft near the aerials.

6.11 Fish processing equipment

6.11.1 Arrangement of fish processing equipment should ensure free access for inspection, operation and sanitary treatment of the equipment. Working areas in way of processing equipment should be not less than 750 mm wide.

6.11.2 Materials used to insulate fish processing equipment, including piping, should be non-combustible, durable and stable under conditions of vibration and should not have an external surface temperature harmful to personnel on contact. The insulation should be securely fastened.

6.11.3 Machinery and installations operating under pressure should comply with the requirements of the Competent Authority.

6.11.4 Machinery and other installations from which vapour, gas, dust or other harmful substance may readily escape or be emitted during operation should be fitted with exhaust devices. The suction ends of these devices should be located, as near as possible, to the sources of vapour, gas, dust or other harmful substance and the piping should be so arranged that discharged products will not constitute a hazard to personnel.

6.11.5 Where several conveyors are working in one line, emergency switches should be provided at intervals of not more than 10 m for stopping all conveyors working in the line. Where the length of the conveyors is 15 m or more, sound or light signals should be provided for giving warning of when the conveyor starts.

6.11.6 Dampers, cocks, valves and other stopping devices should be positioned so that they are readily accessible and safe for operation.

6.11.7 Machinery and equipment in working spaces should be fitted on strong and rigid foundations securely connected to the vessel's structure.

6.11.8 Moving parts of machinery and other equipment or installations, as well as gear wheels, which may present a hazard should be adequately guarded.

6.11.9 Machinery and installations which require routine servicing at a height of more than 2 m should be equipped with platforms of 600 mm in width and guarded with rails not less than 1 m in height.

6.11.10 Fish processing equipment operating with water should be provided with effective drainage systems, having regard to their extra susceptibility to clogging.

6.11.11 Adequate drainage should be provided to prevent the accumulation of water in enclosed spaces as a consequence of fish handling or fish processing.

6.11.12 Loading and unloading devices for fish processing machinery and equipment should be arranged at a safe and convenient height for operation.

6.11.13 Steam or vapour outlets from machinery and equipment such as liver boilers should be arranged as high as possible. Outlet pipes should be at least 50 mm in diameter and lead into open air. Vapour from outlets should not obscure visibility.

6.11.14 Filling openings of machinery and other equipment, such as liver or fish oil boilers, should be within easy reach of personnel. Such openings should be fitted with lids with suitable means of closing so as to prevent steam, hot water or vapour escaping into the working space. The lids should also be counterbalanced or provided with other safe means of securing the lid in the open position.

6.12 Medicine chest, radio-medical services and hospital accommodation

6.12.1 First-aid equipment and instructions as required by the Competent Authorities should be provided in all fishing vessels. International standards relating to first aid at sea laid down in the *International Medical Guide for Ships*, prepared by the International Labour Organization, the International Maritime Organization and the World Health Organization, may serve as a guide. In addition, in recent years regional guidelines have also been developed.*

---

* Refer to EU Council Directive 92/29/EEC on the minimum safety and health requirements for improved medical treatment on board vessels.

6.12.2 Fishing vessels should carry an appropriate medical guide or instructions. The medical guide or instructions should be illustrated, should explain how the medical supplies are to be used and should be designed to enable persons other than a doctor to care for the sick or injured on board both with and, if necessary, without medical advice by radio or satellite communication.

6.12.3 The medicine chest should contain equipment and medical supplies suitable for the expected service of the vessel (e.g. unlimited trips; trips of less than a certain distance from the nearest port with adequate medical equipment; service in harbours and very close to shore).

6.12.4 The Competent Authority should establish requirements for the periodic replacement of medicines to ensure they are not outdated and are appropriate to any changes in the operational requirements of the vessel (e.g. change in geographic location).

6.12.5 Appropriate instructions and equipment should be provided to enable appropriate fishing vessel personnel to consult effectively with radio-medical services ashore.

6.12.6 Appropriate hospital accommodation should be provided in accordance with international instruments.

6.12.7 Instructions and equipment necessary for safe medical evacuation by vessel, helicopter or other means should be carried on board.

6.12.8 Generally, all instructions should be in a language understood by the crew. Where possible, illustrations should be used to facilitate ease of understanding and communication.

6.13 Miscellaneous

6.13.1 Protective clothing and safety working equipment such as gloves, goggles, ear protectors, respirators, safety helmets, special footwear, and/or other apparel, oilskins, explosive gas and oxygen sufficiency indicators, etc. should be provided, as appropriate, to prevent injury or illness to personnel. The protective clothing and, in particular, oilskins should have a highly visible colour, be reflectorized, and fit as closely to the body as possible. The protective clothing for crew members working on deck should be capable of supporting the wearer in the water in the event of being washed overboard. A buoyancy garment or a self-inflating working lifejacket could be used for this purpose.

6.13.2 Pound boards should be so constructed that they can be locked in position when in use and should not hamper the discharge of shipped water.

6.13.3 Effective lightning protectors should be fitted to all wooden masts or topmasts. In vessels built of steel, it is sufficient to fit spikes on steel masts. In vessels constructed of non-conductive material, the lightning conduc-

Chapter VI

tors should be connected to a copper plate fixed to the vessel's hull well below the waterline.

6.13.4 In designing and installing new machinery and equipment in vessels, measures should be taken to reduce the effect of noise and vibration upon personnel to levels satisfactory to the Competent Authority.

6.13.5 Excessive and harmful noise and vibration should, as far as possible, be eliminated. When harmful noise cannot be eliminated, ear protectors should be available to personnel.

6.13.6 An explosive gas detector and a detector or detectors to test for the leakage of refrigerant or refrigerants should be provided on board.

6.13.7 The Competent Authority should ensure that fishing vessels that carry cargo and/or fishing equipment on deck and/or atop deckhouses carry on board clear instructions in relation to:

.1 the provisions in the stability booklet covering conditions of loading at various freeboards;

.2 permitted loading conditions relative to weather conditions;

.3 ensuring that cargo/fishing gear is not stowed in a manner that would obscure view from the bridge or obscure navigation lights and signals; and

.4 ensuring that access to and the operation of essential equipment and machinery is not impeded.

Chapter VII
Life-saving appliances and arrangements

Part A
General

7.1 Definitions

7.1.1 *Float-free launching* is that method of launching a survival craft whereby the craft is automatically released from a sinking vessel and is ready for use.

7.1.2 *Free-fall launching* is that method of launching a survival craft whereby the craft with its complement of persons and equipment on board is released and allowed to fall into the sea without any restraining apparatus.

7.1.3 *Inflatable appliance* is an appliance which depends upon non-rigid, gas-filled chambers for buoyancy and which is normally kept uninflated until ready for use.

7.1.4 *Inflated appliance* is an appliance which depends upon non-rigid, gas-filled chambers for buoyancy and which is kept inflated and ready for use at all times.

7.1.5 *Launching appliance or arrangement* is a means of transferring a survival craft or rescue boat from its stowed position safely to water.

7.1.6 *Novel life-saving appliance or arrangement* is a life-saving appliance or arrangement which embodies new features not fully covered by the provisions of this chapter but which provides an equal or higher standard of safety.

7.1.7 *Rescue boat* is a boat designed to rescue persons in distress and to marshal survival craft.*

---

* Refer to Life-Saving Appliances Code (LSA Code), adopted by the Organization by resolution MSC.48(66).

7.1.8 *Retro-reflective material* is a material which reflects in the opposite direction a beam of light directed on it.*

7.1.9 *Survival craft* is a craft capable of sustaining the lives of persons in distress from the time of abandoning the vessel.*

7.2 Evaluation, testing and approval of life-saving appliances and arrangements†

7.2.1 Except as provided in 7.2.5 and 7.2.6, life-saving appliances and arrangements required by this chapter should be approved by the Competent Authority.

7.2.2 Before giving approval to life-saving appliances and arrangements, the Competent Authority should ensure that such life-saving appliances and arrangements:

.1 are tested, to confirm that they comply with the requirements of this chapter, in accordance with the recommendations of the Organization; or

.2 have successfully undergone, to the satisfaction of the Competent Authority, tests which are substantially equivalent to those specified in those recommendations.

7.2.3 Before giving approval to novel life-saving appliances or arrangements, the Competent Authority should ensure that such appliances or arrangements:

.1 provide safety standards at least equivalent to the requirements of this chapter and the applicable provisions of the Protocol and have been evaluated and tested in accordance with the recommendations of the Organization; or

.2 have successfully undergone, to the satisfaction of the Competent Authority, evaluation and tests which are substantially equivalent to those recommendations.

7.2.4 Procedures adopted by the Competent Authority for approval should also include the conditions whereby approval would continue or would be withdrawn.

7.2.5 Before accepting life-saving appliances and arrangements that have not been previously approved by the Competent Authority, the Competent Authority should be satisfied that life-saving appliances and arrangements

---

* Refer to Life-Saving Appliances Code (LSA Code), adopted by the Organization by resolution MSC.48(66).

† Refer to the Revised Recommendations on the testing of life-saving appliances, adopted by the Organization by resolution MSC.81(70), as revised, and the Code of practice for the evaluation, testing and acceptance of prototype novel life-saving appliances and arrangements, adopted by the Organization by resolution A.520(13).

Life-saving appliances and arrangements

comply with the requirements of this chapter and the applicable provisions of the Protocol.

7.2.6 Notwithstanding provisions of the above paragraphs, alternative life-saving appliances allowed to be installed on board instead of life-saving appliances for which specifications are included in applicable provisions of the Protocol should be to the satisfaction of the Competent Authority.

7.3 Production tests

The Competent Authority should require life-saving appliances to be subjected to such production tests as are necessary to ensure that the life-saving appliances are manufactured to the same standard as the approved prototype.

Part B
Vessel requirements

7.4 Number and types of survival craft

7.4.1 Every vessel should be provided with survival craft of sufficient aggregate capacity to accommodate at least 200% of the total number of persons on board. There should be a sufficient number of survival craft to accommodate at least the total number of persons on board, which should be capable of being launched from either side of the vessel. The Competent Authority may admit a relaxation to this paragraph, taking into account the vessel's navigation area, operational condition, size of the vessel, and may permit to provide vessels with survival craft of sufficient aggregate capacity to accommodate at least the total number of persons on board.

7.4.2 Survival craft should comply with the applicable provisions of the Protocol. Alternatively, the Competent Authority may permit vessels to carry other types of approved survival craft, taking into account the vessel's navigational and operational condition.

7.4.3 The vessel should be provided with a rescue boat. A survival craft can be used as a rescue boat provided it is suitable for this purpose. If a Competent Authority admits that the vessel itself can be used as a rescue boat, and where means for rescuing a person overboard are provided on the vessel, a rescue boat is not necessary. A rescue boat should be capable of being easily launched by a minimum number of crew members, easily propelled and highly manoeuvrable and adequate for rescuing a person overboard. Normally, only rigid boats are considered suitable as a rescue craft, but permanently inflated rubber boats of strong abrasion-resistant construction with subdivided buoyancy may be accepted as rescue boats.

7.5 Availability and stowage of survival craft

7.5.1 Survival craft should:
- .1 be readily available in case of emergency;
- .2 be capable of being launched safely and rapidly under the conditions required by the applicable provisions of the Protocol;
- .3 be so stowed that:
  - .3.1 the marshalling of persons at the embarkation deck is not impeded;
  - .3.2 their prompt handling is not impeded;
  - .3.3 embarkation can be effected rapidly and in good order; and
  - .3.4 the operation of any other survival craft is not interfered with.

7.5.2 Survival craft and launching appliances should be in working order and available for immediate use before the vessel leaves port and kept so at all times when at sea.

7.5.3 Survival craft should be stowed to the satisfaction of the Competent Authority.

7.5.4 Every lifeboat should be attached to a separate set of davits or approved launching appliance.

7.5.5 Survival craft should be positioned as close to accommodation and service spaces as possible, stowed in suitable positions to ensure safe launching, with particular regard to clearance from the propeller. Lifeboats for lowering down the vessel's side should be stowed with regard to steeply overhanging portions of the hull, so ensuring, as far as practicable, that they can be launched down the straight side of the vessel. If positioned forward, they should be stowed abaft the collision bulkhead in a sheltered position and, in this respect, the Competent Authority should give special consideration to the strength of the davits.

7.5.6 The liferafts should be so stowed as to be readily available in case of emergency in such a manner as to permit them to float free from their stowage, inflate and break free from the vessel in the event of its sinking. However, davit-launched liferafts need not float free.

7.5.7 Lashings, if used, should be fitted with an automatic (hydrostatic) release system of an approved type.

7.5.8 All survival craft should be marked with the same registration or other identification marks as used for the vessel as referred to in 7.16.

7.5.9 The Competent Authority, if it is satisfied that the constructional features of the vessel and the method of fishing operation may render it unreasonable and impractical to apply particular provisions of this paragraph, may accept relaxations from such provisions, provided that the vessel is fitted with alternative launching and recovering arrangements adequate for the service for which it is intended.

7.6 Embarkation into survival craft

Suitable arrangements should be made for embarkation into the survival craft, which should include:

.1 at least one ladder, or other approved means, on each side of the vessel to afford access to the survival craft when waterborne except where the Competent Authority is satisfied that the distance from the point of embarkation to the waterborne survival craft is such that a ladder is unnecessary;

.2 means for illuminating the stowage position of survival craft and their launching appliances during preparation for and the process of launching, and also for illuminating the water into which the survival craft are launched until the process of launching is completed, the power for which to be supplied from the emergency source required by section 4.16;

.3 arrangements for warning all persons on board that the vessel is about to be abandoned; and

.4 means for preventing any discharge of water into the survival craft.

7.7 Lifejackets

7.7.1 For every person on board, a lifejacket of an approved type should be carried. Lifejackets should comply with the provisions of the Recommendations for testing lifejackets, reproduced at annex V to this part of the Code.

7.7.2 Lifejackets should be so placed as to be readily accessible and their position should be plainly indicated.

7.8 Immersion suits and thermal protective aids

7.8.1 For vessels operating in areas where low water and air temperature can be expected, an approved immersion suit of an appropriate size should be provided for every person on board.

7.8.2 Where the Competent Authority considers that water and air temperatures in the area of operations of the vessel warrant immersion suits with inherent insulation, these suits should be provided for every person on board.

7.8.3 Immersion suits should be placed as to be readily accessible and their position should be clearly indicated.

Chapter VII: part B

7.9 Lifebuoys

7.9.1 At least four lifebuoys should be provided.

7.9.2 At least half of the number of lifebuoys referred to in 7.9.1 should be provided with self-igniting lights.

7.9.3 At least one of the lifebuoys provided with self-igniting lights in accordance with 7.9.2 should be provided with self-activating smoke signals.

7.9.4 At least one lifebuoy on each side of the vessel should be fitted with a buoyant lifeline equal in length to not less than 30 m. Such lifebuoys should not have self-igniting lights.

7.9.5 All lifebuoys should be so placed as to be readily accessible to the persons on board and should always be capable of being rapidly cast loose and should not be permanently secured in any way.

7.9.6 All lifebuoys should be in a bright contrasting colour to the sea and marked with the same registration or other identification marks as used for the vessel as referred to in 7.16.

7.10 Line-throwing appliances

For every vessel a line-throwing appliance of a type approved by the Competent Authority should be provided, being capable of carrying with reasonable accuracy a line not less than 230 m in length.

7.11 Distress signals

7.11.1 Every vessel should be provided, to the satisfaction of the Competent Authority, with means of making effective distress signals by day and by night, including at least 6 rocket parachute flares.

7.11.2 Distress signals should be of an approved type. They should be so placed as to be readily accessible and their position should be plainly indicated.

7.12 Radio life-saving appliances

At least two two-way VHF radiotelephone apparatus should be provided on every vessel. Such apparatus should conform to performance standards not inferior to those adopted by the Competent Authority, having regard to those adopted by the Organization. If a fixed two-way VHF radiotelephone apparatus is fitted in a survival craft, it should conform to performance standards not inferior to those adopted by the Competent Authority, having regard to those adopted by the Organization.

7.13 Radar transponders*

At least one radar transponder should be carried on every vessel. Such radar transponders should conform to performance standards not inferior to those adopted by the Competent Authority, having regard to those adopted by the Organization. It should be stowed in such a location that it can be rapidly placed in any survival craft.

7.14 Retro-reflective materials on life-saving appliances

All survival craft, rescue boats, lifejackets, immersion suits and lifebuoys should be fitted with retro-reflective material in accordance with the recommendations of the Organization.

7.15 Operational readiness, maintenance and inspections

Operational readiness

7.15.1 Before the vessel leaves port and at all times during the voyage, all life-saving appliances should be in working order and ready for immediate use.

Maintenance

7.15.2 Instructions for on-board maintenance of life-saving appliances approved by the Competent Authority should be provided and maintenance should be carried out accordingly.

7.15.3 The Competent Authority may accept, in lieu of the instructions required by 7.15.2, a shipboard planned maintenance programme.

Maintenance of falls

7.15.4 Falls used in launching should be turned end for end at intervals of not more than 30 months and be renewed when necessary due to deterioration of the falls or at intervals of not more than five years, whichever is the earlier.

Spares and repair equipment

7.15.5 Spares and repair equipment should be provided for life-saving appliances and their components which are subject to excessive wear or consumption and need to be replaced regularly.

---

* Refer to the Performance standards for survival craft radar transponders for use in search and rescue operations, adopted by the Organization by resolution A.802(19).

Chapter VII: part B

Weekly inspection

7.15.6 The following tests and inspections should be carried out weekly:

.1 all survival craft and launching appliances should be visually inspected to ensure that they are ready for use;

.2 all engines in lifeboats should be run ahead and astern for a total period of not less than 3 min provided the ambient temperature is above the minimum temperature required for starting the engine; and

.3 the general emergency alarm system should be tested.

Monthly inspections

7.15.7 Inspection of the life-saving appliances, including lifeboat equipment, should be carried out monthly, using a checklist to ensure that they are complete and in good order. A report of the inspection should be entered in the log-book.

Servicing of inflatable liferafts and inflatable lifejackets

7.15.8 Every inflatable liferaft and inflatable lifejacket should be serviced:

.1 at intervals not exceeding 12 months. However, in cases where it appears proper and reasonable, the Competent Authority may extend this period to 17 months; and

.2 at an approved servicing station which is competent to service them, maintains proper servicing facilities and uses only properly trained personnel.

Periodic servicing of hydrostatic release units

7.15.9 Hydrostatic release units should be serviced:

.1 at intervals not exceeding 12 months. However, in cases where it appears proper and reasonable, the Competent Authority may extend this period to 17 months; and

.2 at a servicing station which is competent to service them, maintains proper servicing facilities and uses only properly trained personnel.

7.15.10 In cases of vessels where the nature of fishing operations may cause difficulty for compliance with the requirements of 7.15.8 and 7.15.9, the Competent Authority may allow the extension of the service intervals to 24 months, provided that the Competent Authority is satisfied that such appliances are so manufactured and arranged that they will remain in satisfactory condition until the next period of servicing.

7.16 Miscellaneous

To facilitate aerial rescue operations, wheelhouse tops or other prominent horizontal surfaces should be painted in a highly visible colour and should bear the vessel's registration or other identification marks in letters and/or numerals in contrasting colours. Similar marks on the sides of the wheelhouse would also facilitate search and identification by high-speed aircraft.*

Part C
Life-saving appliance requirements

Subject to 1.1.3, part C of chapter VII of the Torremolinos International Convention for the Safety of Fishing Vessels, 1977, as modified by the Torremolinos Protocol of 1993 relating thereto should be used as guidance for requirements for life-saving appliances.

---

* Marking of fishing vessels and fishing gear for identification should be in accordance with uniform and internationally-recognizable vessel and gear marking systems, such as the Food and Agriculture Organization of the United Nations Standard Specifications for the Marking and Identification of Fishing Vessels. Refer to the FAO Technical Guidelines for Responsible Fisheries – No 1 Fishing Operations. (ISBN 92-5-103914-3) and MSC/Circ.572.

Chapter VIII
Emergency procedures, musters and drills

8.1 General emergency alarm system, muster list and emergency instructions

8.1.1 The general emergency alarm system should be capable of sounding the general alarm signal consisting of seven or more short blasts followed by one long blast on the vessel's whistle or siren and, additionally, on an electrically operated bell or klaxon or other equivalent warning system which should be powered from the vessel's main supply and the emergency source of electrical power required by 4.16.

8.1.2 All vessels should be provided with clear instructions for each crew member which should be followed in case of emergency.

8.1.3 The muster list should be posted up in several parts of the vessel and, in particular, in the wheelhouse, the engine-room and in the crew accommodation and should include the information specified in the following paragraphs.

8.1.4 The muster list should specify details of the general alarm signal prescribed by 8.1.1 and also the action to be taken by the crew when this alarm is sounded. The muster list should also specify how the order to abandon ship will be given.

8.1.5 The muster list should show the duties assigned to the different members of the crew, including:

- .1 closing of watertight doors, fire doors, valves, scuppers, overboard shoots, sidescuttles, skylights, portholes and other similar openings in the vessel;
- .2 equipping the survival craft and other life-saving appliances;
- .3 preparation and launching of survival craft;
- .4 general preparation of other life-saving appliances;
- .5 use of communication equipment; and
- .6 manning of fire parties assigned to deal with fires.

8.1.6 In vessels of less than 45 m in length, the Competent Authority may permit relaxation of the requirements of 8.1.5 if satisfied that, due to the small number of crew members, no muster list is necessary.

8.1.7 The muster list should specify which officers are assigned to ensure that the life-saving and fire appliances are maintained in good condition and are ready for immediate use.

8.1.8 The muster list should specify substitutes for key persons who may become disabled, taking into account that different emergencies may call for different actions.

8.1.9 The muster list should be prepared before the vessel proceeds to sea. After the muster list has been prepared, if any change takes place in the crew which necessitates an alteration in the muster list, the skipper should either revise the list or prepare a new list.

8.2 Abandon-ship training and drills

Practice musters and drills

8.2.1 Each member of the crew should participate in at least one abandon-ship drill and one fire drill every month. However, on vessels less than 45 m in length, the Competent Authority may modify this requirement, provided that at least one abandon-ship and one fire drill is held at least every three months. The drills of the crew should take place within 24 h of the vessel leaving a port if more than 25% of the crew have not participated in abandon-ship and fire drills on board that particular vessel in the previous muster. The Competent Authority may accept other arrangements that are at least equivalent for those classes of vessel for which this is impracticable.

8.2.2 Each abandon-ship drill should include:

.1 summoning of crew to muster stations with the general emergency alarm and ensuring that they are made aware of the order to abandon ship specified in the muster list;

.2 reporting to stations and preparing for the duties described in the muster list;

.3 checking that crew are suitably dressed;

.4 checking that lifejackets are correctly donned;

.5 lowering of at least one lifeboat after any necessary preparation for launching;

.6 starting and operating the lifeboat engine;

.7 operation of davits used for launching liferafts.

8.2.3 Each fire drill should include:

.1 reporting to stations and preparing for the duties described in the fire muster list;

Chapter VIII

.2 starting of a fire pump, using at least the two required jets of water to show that the system is in proper working order;

.3 checking of fireman's outfit and other personal rescue equipment;

.4 checking of relevant communication equipment;

.5 checking the operation of watertight doors, fire doors, fire dampers and means of escape; and

.6 checking the necessary arrangements for subsequent abandoning of the vessel.

8.2.4 Different lifeboats should, as far as practicable, be lowered in compliance with the requirements of 8.2.2.5 at successive drills.

8.2.5 Drills should, as far as practicable, be conducted as if there were an actual emergency.

8.2.6 Each lifeboat should be launched with its assigned operating crew aboard and manoeuvred in the water at least once every 3 months during an abandon-ship drill.

8.2.7 As far as is reasonable and practicable, rescue boats other than lifeboats which are also rescue boats should be launched each month with their assigned crew aboard and manoeuvred in the water. In all cases, this requirement should be complied with at least once every 3 months.

8.2.8 If lifeboat and rescue boat launching drills are carried out with the vessel making headway, such drills should, because of the dangers involved, be practised in sheltered waters only and under the supervision of an officer experienced in such drills.

8.2.9 Emergency lighting for mustering and abandonment should be tested at each abandon-ship drill.

8.2.10 The drills may be adjusted according to the relevant equipment required by those regulations. However, if equipment is carried on a voluntary basis, it should be used in the drills and the drills should be adjusted accordingly.

On-board training and instructions

8.2.11 On-board training in the use of the vessel's life-saving appliances, including survival craft equipment, should be given as soon as possible but not later than 2 weeks after a crew member joins the vessel. However, if the crew member is on a regularly scheduled rotating assignment to the vessel, such training should be given not later than 2 weeks after the time of first joining the vessel.

8.2.12 Instructions in the use of the vessel's life-saving appliances and in survival at sea should be given at the same intervals as the drills. Individual instruction may cover different parts of the vessel's life-saving system, but all the vessel's life-saving equipment and appliances should be covered within any period of 2 months. Each member of the crew should be given instructions which should include, but not necessarily be limited to:

    .1   operation and use of the vessel's inflatable liferafts, including precautions concerning nailed shoes and other sharp objects;

    .2   problems of hypothermia, first-aid treatment for hypothermia and other appropriate first-aid procedures; and

    .3   special instructions necessary for use of the vessel's life-saving appliances in severe weather and severe sea conditions.

8.2.13 On-board training in the use of davit-launched liferafts should take place at intervals of not more than 4 months on every vessel fitted with such appliances. Whenever practicable, this should include the inflation and lowering of a liferaft. This liferaft may be a special liferaft intended for training purposes only, which is not part of the vessel's life-saving equipment; such a special liferaft should be conspicuously marked.

Records

8.2.14 The date when musters are held, details of abandon-ship drills and fire drills, drills of other life-saving appliances and on-board training should be recorded in such log-book as may be prescribed by the Competent Authority. If a full muster, drill or training session is not held at the appointed time, an entry should be made in the log-book stating the circumstances and the extent of the muster, drill or training session held.

Training manual

8.2.15 A training manual should be provided in each crew mess room and recreation room or in each crew cabin. The training manual, which may comprise several volumes, should contain instructions and information, in easily understood terms illustrated wherever possible, on the life-saving appliances provided in the vessel and on the best methods of survival. Any part of such information may be provided in the form of audio-visual aids in lieu of the manual. The following should be explained in detail:

    .1   donning of lifejackets and immersion suits, as appropriate;

    .2   muster at the assigned stations;

    .3   boarding, launching and clearing the survival craft and rescue boats;

    .4   method of launching from within the survival craft;

    .5   release from launching appliances;

    .6   methods and use of devices for protection in launching areas, where appropriate;

    .7   illumination in launching areas;

    .8   use of all survival equipment;

    .9   use of all detection equipment;

- .10 with the assistance of illustrations, the use of radio life-saving appliances;
- .11 use of drogues;
- .12 use of engine and accessories;
- .13 recovery of survival craft and rescue boats, including stowage and securing;
- .14 hazards of exposure and the need for warm clothing;
- .15 best use of the survival craft facilities in order to survive;
- .16 methods of retrieval, including the use of helicopter rescue gear (slings, baskets, stretchers), breeches-buoy and shore life-saving apparatus and vessel's line-throwing apparatus;
- .17 all other functions contained in the muster list and emergency instructions; and
- .18 instructions for emergency repair of the life-saving appliances.

8.2.16 On vessels of less than 45 m in length, the Competent Authority may permit relaxation of the requirements of 8.2.15 above. However, appropriate safety information should be carried on board.

8.3 Training in emergency procedures

The Competent Authority should take such measures as it may deem necessary to ensure that crews are adequately trained in their duties in the event of emergencies. Such training should include, as appropriate:

- .1 types of emergencies which may occur, such as collisions, fire and foundering;
- .2 types of life-saving appliances normally carried on vessels;
- .3 need to adhere to the principles of survival;
- .4 value of training and drills;
- .5 need to be ready for any emergency and to be constantly aware of:
    - .5.1 the information in the muster list, in particular:
        - .1 each crew member's specific duties in any emergency;
        - .2 each crew member's own survival station; and
        - .3 the signals calling the crew to their survival craft or fire stations;
    - .5.2 location of each crew member's own and spare lifejackets;
    - .5.3 location of fire alarm controls;
    - .5.4 means of escape;
    - .5.5 consequences of panic;

- .6 actions to be taken in respect to lifting persons from vessels and survival craft by helicopter;
- .7 actions to be taken when called to survival craft stations, including:
- .7.1 putting on suitable clothing;
- .7.2 donning of lifejacket; and
- .7.3 collecting additional protection such as blankets, time permitting;
- .8 actions to be taken when required to abandon ship, such as:
- .8.1 how to board survival craft from vessel and water; and
- .8.2 how to jump into the sea from a height and reduce the risk of injury when entering the water;
- .9 actions to be taken when in the water, such as:
- .9.1 how to survive in circumstances of:
  - .1 fire or oil on the water;
  - .2 cold conditions; and
  - .3 shark-infested waters;
- .9.2 how to right a capsized survival craft;
- .10 actions to be taken when aboard a survival craft, such as:
- .10.1 getting the survival craft quickly clear of the vessel;
- .10.2 protection against cold or extreme heat;
- .10.3 using a drogue or sea anchor;
- .10.4 keeping a look-out;
- .10.5 recovering and caring for survivors;
- .10.6 facilitating detection by others;
- .10.7 checking equipment available for use in the survival craft and using it correctly; and
- .10.8 remaining, so far as possible, in the vicinity;
- .11 main dangers to survivors and the general principles of survival, including:
- .11.1 precautions to be taken in cold climates;
- .11.2 precautions to be taken in tropical climates;
- .11.3 exposure to sun, wind, rain and sea;
- .11.4 importance of wearing suitable clothing;
- .11.5 protective measures in survival craft;
- .11.6 effects of immersion in the water and of hypothermia;
- .11.7 importance of preserving body fluids;
- .11.8 protection against seasickness;

- .11.9 proper use of fresh water and food;
- .11.10 effects of drinking seawater;
- .11.11 means available for facilitating detection by others; and
- .11.12 importance of maintaining morale;
- .12 actions to be taken in respect to fire fighting:
- .12.1 the use of fire hoses with different nozzles;
- .12.2 the use of fire extinguishers;
- .12.3 knowledge of the location of fire doors; and
- .12.4 the use of breathing apparatus.

Chapter IX
Radiocommunications

Part A
General

9.1 Application and definitions

9.1.1 This chapter should apply to new and existing fishing vessels.

9.1.2 No provision in this chapter should prevent the use by any vessel, survival craft or person in distress of any means at its disposal to attract attention, make known its position and obtain help.

9.1.3 For the purpose of this chapter, the following terms should have the meanings defined below:

9.1.3.1 *Bridge-to-bridge communications* means safety communications between vessels from the position from which the vessels are normally navigated.

9.1.3.2 *Continuous watch* means that the radio watch concerned should not be interrupted other than for brief intervals when the vessel's receiving capability is impaired or blocked by its own communications or when the facilities are under periodical maintenance or checks.

9.1.3.3 *Digital selective calling (DSC)* means a technique using digital codes which enables a radio station to establish contact with, and transfer information to, another station or group of stations, and complying with the relevant recommendations of the International Radio Consultative Committee (CCIR).

9.1.3.4 *Direct printing telegraphy* means automated telegraphy techniques which comply with the relevant recommendations of the CCIR.

9.1.3.5 *General radiocommunications* means operational and public correspondence traffic, other than distress, urgency and safety messages, conducted by radio.

9.1.3.6 *Inmarsat* means the Organization established by the Convention on the International Maritime Satellite Organization adopted on 3 September 1976.

9.1.3.7 *International NAVTEX Service* means the co-ordinated broadcast and automatic reception on 518 kHz of maritime safety information by means of narrow-band direct-printing telegraphy using the English language.*

9.1.3.8 *Locating* means the finding of ships, vessels, aircraft, units or persons in distress.

9.1.3.9 *Maritime safety information* means navigational and meteorological warnings, meteorological forecasts and other urgent safety-related messages broadcast to vessels.

9.1.3.10 *Polar orbiting satellite service* means a service which is based on polar-orbiting satellites which receive and relay distress alerts from satellite emergency position-indicating radio beacons (satellite EPIRBs) and which provides their position.

9.1.3.11 *Radio Regulations* means the Radio Regulations annexed to, or regarded as being annexed to, the most recent International Telecommunication Convention which is in force at any time.

9.1.3.12 *Sea area A1* means an area within the radiotelephone coverage of at least one VHF coast station in which continuous DSC alerting is available, as may be defined by a Party.†

9.1.3.13 *Sea area A2* means an area, excluding sea area A1, within the radiotelephone coverage of at least one MF coast station in which continuous DSC alerting is available, as may be defined by a Party.*

9.1.3.14 *Sea area A3* means an area, excluding sea areas A1 and A2, within the coverage of an Inmarsat geostationary satellite in which continuous alerting is available.

9.1.3.15 *Sea area A4* means an area outside sea areas A1, A2 and A3.

9.1.4 All other terms and abbreviations which are used in this chapter and which are defined in the Radio Regulations should have the meanings as defined in those Regulations.

9.2 Exemptions

9.2.1 It is highly desirable not to deviate from the requirements of this chapter; nevertheless the Competent Authority may grant partial or

---

* Refer to the *NAVTEX Manual*, approved by the Organization (publication IC951E).
† Refer to resolution A.704(17), Provision of radio services for the global maritime distress and safety system (GMDSS).

conditional exemptions to individual vessels from the requirements of 9.5 to 9.9 provided:

.1 such vessels comply with the functional requirements of 9.3; and

.2 the Competent Authority has taken into account the effect such exemption may have upon the general efficiency of the service for the safety of all ships and vessels.

9.2.2 An exemption may be permitted under paragraph 9.2.1 only:

.1 if the conditions affecting safety are such as to render the full application of 9.5 to 9.9 unreasonable or unnecessary; or

.2 in exceptional circumstances, for a single voyage outside the sea area or sea areas for which the vessel is equipped.

9.2.3 The Competent Authority may exempt vessels operating always together in pairs or in groups from being fully equipped in accordance with the requirements, provided that:

.1 the vessel in command fully complies with the requirements of the actual sea area;

.2 the other vessels in a pair or in groups carry radio equipment sufficient for short-distance distress alert and radiocommunications with the vessel in command, to the satisfaction of the Competent Authority. "Vessels operating in a pair or group" is defined as two or more vessels operating collaboratively within 100 nautical miles of each other except for extremely brief periods; and

.3 this exemption does not apply to EPIRB carriage requirements.

9.3 Functional requirements

Every vessel, while at sea, should be capable:

.1 except as provided in 9.6.1.1 and 9.8.1.4.3, of transmitting ship-to-shore distress alerts by at least two separate and independent means, each using a different radiocommunication service;

.2 of receiving shore-to-ship distress alerts;

.3 of transmitting and receiving ship-to-ship distress alerts;

.4 of transmitting and receiving search and rescue co-ordinating communications;

.5 of transmitting and receiving on-scene communications;

.6 of transmitting and, as required by regulation X/3(6) of the Protocol, receiving signals for locating;

.7 of transmitting and receiving maritime safety information;

Chapter IX: part B

.8 of transmitting and receiving general radiocommunications to and from shore-based radio systems or networks subject to 9.13.7; and

.9 of transmitting and receiving bridge-to-bridge communications.

Part B
Ship requirements

9.4 Radio installations

9.4.1 Every vessel should be provided with radio installations capable of complying with the functional requirements prescribed by 9.3 throughout its intended voyage and, unless relaxed under 9.2, complying with the requirements of 9.5 and, as appropriate for the sea area or areas through which it will pass during its intended voyage, the requirements of either 9.5, 9.6, 9.7 or 9.8.

9.4.2 Every radio installation should:

.1 be so located that no harmful interference of mechanical, electrical or other origin affects its proper use, and so as to ensure electromagnetic compatibility and avoidance of harmful interaction with other equipment and systems;

.2 be so located as to ensure the greatest possible degree of safety and operational availability;

.3 be protected against harmful effects of water, extremes of temperature and other adverse environmental conditions;

.4 be provided with reliable, permanently arranged electrical lighting, independent of the main and emergency sources of electrical power, for the adequate illumination of the radio controls for operating the radio installation; and

.5 be clearly marked with the call sign, the ship station identity and other codes as applicable for the use of the radio installation. This includes the maritime mobile service identities (MMSI).

9.4.3 Control of the VHF radiotelephone channels, required for navigational safety, should be immediately available on the navigation bridge convenient to the conning position and, where necessary, facilities should be available to permit radiocommunications from the wings of the navigation bridge. Portable VHF equipment may be used to meet the latter provision.

Radiocommunications

9.5 Radio equipment – General

9.5.1 Every vessel should be provided with:

.1 a VHF radio installation capable of transmitting and receiving:

.1.1 DSC on the frequency 156.525 MHz (channel 70). It should be possible to initiate the transmission of distress alerts on channel 70 from the position from which the vessel is normally navigated; and

.1.2 radiotelephony on the frequencies 156.300 MHz (channel 6), 156.650 MHz (channel 13) and 156.800 MHz (channel 16);

.2 a VHF DSC watch receiver which may be separate from, or combined with, that required by 9.5.1.1.1;

.3 a radar transponder capable of operating in the 9 GHz band, which:

.3.1 should be so stowed that it can be easily utilized; and

.3.2 may be one of those required by 7.13 for a survival craft;

.4 a receiver capable of receiving International NAVTEX Service broadcasts if the ship is engaged on voyages in any area in which an International NAVTEX Service is provided. However, if a NAVTEX service is not established in the actual area, the Competent Authority may permit vessels to receive navigational warnings and safety messages by other means of reception, to the satisfaction of the Competent Authority;

.5 a radio facility for reception of maritime safety information by the Inmarsat enhanced group calling system, if the vessel is engaged on voyages in any area of Inmarsat coverage but in which a NAVTEX or an alternative service is not provided. However, vessels engaged exclusively on voyages in areas where an HF direct-printing telegraphy maritime safety information service is provided, and fitted with equipment capable of receiving such service, may be exempted from this requirement;

.6 a satellite emergency position-indicating radio beacon (satellite EPIRB) which should be:

.6.1 capable of transmitting a distress alert either through the polar orbiting satellite service operating in the 406 MHz band or, if the vessel is engaged only on voyages within Inmarsat coverage, through the Inmarsat geostationary satellite service operating in the 1.6 GHz band;

.6.2 installed in an easily accessible position;

.6.3 ready to be manually released and capable of being carried by one person into a survival craft;

.6.4 capable of floating free if the vessel sinks and of being automatically activated when afloat; and

.6.5 capable of being activated manually.

9.6 Radio equipment – Sea area A1 or sea areas within the coverage of a VHF coast station (without DSC) operating on a 24 hours a day, 7 days a week basis

9.6.1 In addition to meeting the requirements of 9.5, every vessel engaged on voyages exclusively in sea area A1 should be provided with a radio installation capable of initiating the transmission of ship-to-shore distress alerts from the position from which the vessel is normally navigated, operating either:

.1 on VHF using DSC; this requirement may be fulfilled by the EPIRB prescribed by 9.6.3, either by installing the EPIRB close to, or by remote activation from, the position from which the vessel is normally navigated; or

.2 through the polar orbiting satellite service on 406 MHz; this requirement may be fulfilled by the satellite EPIRB required by 9.5.1.6, either by installing the satellite EPIRB close to, or by remote activation from, the position from which the vessel is normally navigated; or

.3 if the vessel is engaged on voyages within coverage of MF coast stations equipped with DSC, on MF using DSC; or

.4 on HF using DSC; or

.5 through the Inmarsat geostationary satellite service; this requirement may be fulfilled by:

.5.1 an Inmarsat ship earth station; or

.5.2 the satellite EPIRB required by 9.5.1.6, either by installing the satellite EPIRB close to, or by remote activation from, the position from which the vessel is normally navigated.

9.6.2 The VHF radio installation, required by 9.5.1.1, should also be capable of transmitting and receiving general radiocommunications using radiotelephony.

9.6.3 Vessels engaged on voyages exclusively in sea area A1 may carry, in lieu of the satellite EPIRB required by 9.5.1.6, an EPIRB which should be:

.1 capable of transmitting a distress alert using DSC on VHF channel 70 and providing for locating by means of a radar transponder operating in the 9 GHz band;

.2 installed in an easily accessible position;

.3 ready to be manually released and capable of being carried by one person into a survival craft;

.4 capable of floating free if the vessel sinks and being automatically activated when afloat; and

.5 capable of being activated manually.

Radiocommunications

9.7 Radio equipment – Sea areas A1 and A2 or sea areas within the coverage of an MF coast station (without DSC) providing a continuous watch on 2182 kHz as well as a continuously operating VHF station

9.7.1 In addition to meeting the requirements of 9.5 and 9.6, every vessel engaged on voyages beyond sea area A1, but remaining within sea area A2, should be provided with:

.1 an MF radio installation capable of transmitting and receiving, for distress and safety purposes, on the frequencies:

.1.1 2187.5 kHz using DSC; and

.1.2 2182 kHz using radiotelephony;

.2 a radio installation capable of maintaining a continuous DSC watch on the frequency 2187.5 kHz which may be separate from or combined with that required by 9.7.1.1; and

.3 means of initiating the transmission of ship-to-shore distress alerts by a radio service other than MF operating either:

.3.1 through the polar orbiting satellite service on 406 MHz; this requirement may be fulfilled by the satellite EPIRB required by 9.5.1.6, either by installing the satellite EPIRB close to, or by remote activation from, the position from which the vessel is normally navigated; or

.3.2 on HF using DSC; or

.3.3 through the Inmarsat geostationary satellite service; this requirement may be fulfilled by an Inmarsat ship earth station or by the satellite EPIRB, required by 9.5.1.6, either by installing the satellite EPIRB close to, or by remote activation from, the position from which the vessel is normally navigated.

9.7.2 It should be possible to initiate transmission of distress alerts by the radio installations specified in 9.7.1.1 and 9.7.1.3 from the position from which the vessel is normally navigated.

9.7.3 The vessel should, in addition, be capable of transmitting and receiving general radiocommunications using radiotelephony or direct-printing telegraphy by either:

.1 a radio installation operating on working frequencies in the bands between 1605 kHz and 4000 kHz or between 4000 kHz and 27,500 kHz. This requirement may be fulfilled by the addition of this capability in the equipment required by 9.7.1.1; or

.2 an Inmarsat ship earth station.

Chapter IX: part B

9.7.4 If the vessel is operating exclusively within the radiotelephone coverage of at least one MF coast station in which continuous DSC alerting is not available, but is providing a continuous watch on 2182 kHz, the vessel need not to be equipped with the DSC functions mentioned above in 9.5.1.1, 9.5.1.2, 9.7.1.1 to 9.7.1.3.

9.8 Radio equipment – Sea areas A1, A2 and A3

9.8.1 In addition to meeting the requirements of 9.5, 9.6 and 9.7, every vessel engaged on voyages beyond sea areas A1 and A2, but remaining within sea area A3, should, if it does not comply with the requirements of 9.8.2, be provided with:

.1   an Inmarsat ship earth station capable of:

.1.1   transmitting and receiving distress and safety communications using either radiotelephony or direct-printing telegraphy;

.1.2   initiating and receiving distress priority calls;

.1.3   maintaining watch for shore-to-ship distress alerts, including those directed to specifically defined geographical areas;

.1.4   transmitting and receiving general radiocommunications, using either radiotelephony or direct-printing telegraphy; and

.2   an MF radio installation capable of transmitting and receiving, for distress and safety purposes, on the frequencies:

.2.1   2187.5 kHz using DSC; and

.2.2   2182 kHz using radiotelephony, and

.3   a radio installation capable of maintaining a continuous DSC watch on the frequency 2187.5 kHz which may be separate from or combined with that required by 9.8.1.2.1; and

.4   means of initiating the transmission of ship-to-shore distress alerts by a radio service operating either:

.4.1   through the polar orbiting satellite service on 406 MHz; this requirement may be fulfilled by the satellite EPIRB, required by 9.5.1.6, either by installing the satellite EPIRB close to, or by remote activation from, the position from which the vessel is normally navigated; or

.4.2   on HF using DSC; or

.4.3   through the Inmarsat geostationary satellite service, by an additional ship earth station or by the satellite EPIRB required by 9.5.1.6, either by installing the satellite EPIRB close to, or by remote activation from, the position from which the vessel is normally navigated.

9.8.2 In addition to meeting the requirements of 9.5, 9.6, and 9.7, every vessel engaged on voyages beyond sea areas A1 and A2, but remaining within sea area A3, should, if it does not comply with the requirements of 9.8.1, be provided with:

.1 an MF/HF radio installation capable of transmitting and receiving, for distress and safety purposes, on all distress and safety frequencies in the bands between 1609 kHz and 4000 kHz and between 4000 kHz and 27,500 kHz:

.1.1 using DSC; and

.1.2 using radiotelephony;

.2 equipment capable of maintaining DSC watch on 2187.5 kHz, 8414.5 kHz and on at least one of the distress and safety DSC frequencies 4207.5 kHz, 6312 kHz, 12,577 kHz or 16,804.5 kHz; at any time, it should be possible to select any of these DSC distress and safety frequencies. This equipment may be separate from or combined with the equipment required by 9.8.2.1; and

.3 means of initiating the transmission of ship-to-shore distress alerts by a radiocommunication service other than HF operating either:

.3.1 through the polar orbiting satellite service on 406 MHz; this requirement may be fulfilled by the satellite EPIRB, required by 9.5.1.6, either by installing the satellite EPIRB close to, or by remote activation from, the position from which the vessel is normally navigated; or

.3.2 through the Inmarsat geostationary satellite service; this requirement may be fulfilled by an Inmarsat ship earth station or the satellite EPIRB, required by 9.5.1.6, either by installing the satellite EPIRB close to, or by remote activation from, the position from which the vessel is normally navigated.

9.8.3 It should be possible to initiate transmission of distress alerts by the radio installations specified in 9.8.1.1, 9.8.1.2, 9.8.1.4, 9.8.2.1 and 9.8.2.3 from the position from which the vessel is normally navigated.

9.9 Additional note on relaxation – Sea area A3

Notwithstanding the provisions of 9.5, the Competent Authority may permit exemption of the provisions of 9.5.1.1 and 9.5.1.2 in areas where such shore-based services are not available.

9.10 Watches

9.10.1 Every vessel, while at sea, should maintain either a continuous watch:

.1 on VHF DSC channel 70, if the vessel, in accordance with the requirements of 9.5.1.2, is fitted with a VHF radio installation;

> .2 on the distress and safety DSC frequency 2187.5 kHz, if the vessel, in accordance with the requirements of 9.7.1.2 or 9.8.1.3, is fitted with an MF radio installation;
>
> .3 on the distress and safety DSC frequencies 2187.5 kHz and 8414.5 kHz and also on at least one of the distress and safety DSC frequencies 4207.5 kHz, 6312 kHz, 12,577 kHz or 16,804.5 kHz, appropriate to the time of day and the geographical position of the vessel, if the vessel, in accordance with the requirements of 9.8.2.2, is fitted with an MF/HF radio installation. This watch may be kept by means of a scanning receiver;
>
> .4 for satellite shore-to-ship distress alerts, if the vessel, in accordance with the requirements of 9.8.1.1, is fitted with an Inmarsat ship earth station; or
>
> .5 on the radiotelephone distress frequency 2182 kHz, if the vessel is operating within the radiotelephone coverage of an MF coast station in which continuous DSC alerting is not available or is not fitted with the MF DSC functions in 9.7.1.1 and 9.7.1.2. This watch should be kept at the position from which the vessel is normally navigated.

9.10.2 Every vessel, while at sea, should maintain a radio watch for broadcasts of maritime safety information on the appropriate frequency or frequencies on which such information is broadcast for the area in which the vessel is navigating.

9.10.3 Every vessel, while at sea, should maintain, when practicable, a continuous listening watch on VHF channel 16.

9.11 Sources of energy

9.11.1 There should be available at all times, while the vessel is at sea, a supply of electrical energy sufficient to operate the radio installations and to charge any batteries used as part of a reserve source or sources of energy for the radio installations.

9.11.2 A reserve source or sources of energy should be provided on every vessel, to the satisfaction of the Competent Authority, to supply radio installations, for the purpose of conducting distress and safety radiocommunications, in the event of failure of the vessel's main and emergency source of electrical power. The reserve source of energy should be capable of simultaneously operating:

> .1 the VHF radio installation in sea area A1;
>
> .2 the VHF radio installation and the MF or MF/HF installation in sea area A2;
>
> .3 the VHF radio installation and the MF or MF/HF installation or the Inmarsat station in sea area A3; and
>
> .4 for a period of at least 3 h.

The reserve source of energy need not supply independent HF and MF radio installation at the same time.

9.11.3 The reserve source or sources of energy should be independent of the propelling power of the vessel and the vessel's electrical system.

9.11.4 The reserve source or sources of energy may be used to supply the electrical lighting required by 9.4.2.4.

9.11.5 Where a reserve source of energy consists of a rechargeable accumulator battery or batteries:

.1 a means of automatically charging such batteries should be provided which should be capable of recharging them to minimum capacity requirements within 10 h; and

.2 the capacity of the battery or batteries should be checked, using an appropriate method, at intervals not exceeding 12 months.

9.12 Performance standards

All equipment to which this chapter applies should be of a type approved by the Competent Authority. Such equipment, except for the domestic radio installation and its ancillary equipment, should conform to appropriate performance standards approved by the Competent Authority, having regard to those adopted by the Organization.

9.13 Maintenance requirements

9.13.1 Equipment should be so designed that the main units can be replaced readily, without elaborate re-calibration or readjustment.

9.13.2 Where applicable, equipment should be so constructed and installed that it is readily accessible for inspection and on-board maintenance purposes.

9.13.3 Adequate information should be provided to enable the equipment to be properly operated and maintained, taking into account the recommendations of the Organization.

9.13.4 Adequate tools and spares should be provided to enable the equipment to be maintained.

9.13.5 The Competent Authority should ensure that radio equipment required by this chapter is maintained to provide the availability of the functional requirements specified in 9.3 and to meet the recommended performance standards of such equipment.

9.13.6 On vessels engaged on voyages in sea area A3, the availability should be ensured by using such methods as duplication of equipment, shore-based maintenance or at-sea electronic maintenance capability, or a combination of these, as may be approved by the Competent Authority.

9.13.7 While all reasonable steps should be taken to maintain the equipment in efficient working order to ensure compliance with all the functional requirements specified in 9.3, malfunction of the equipment for providing the general radiocommunications required by 9.3.8 should not be considered as making a vessel unseaworthy or as a reason for delaying the vessel in ports where repair facilities are not readily available, provided the vessel is capable of performing all distress and safety functions.

9.13.8 Satellite EPIRBs should be annually tested for all aspects of operational efficiency, with special emphasis on checking the emission on operational frequencies, coding and registration. The test may be conducted on board the vessel or at an approved testing station. Satellite EPIRBs should be subject to maintenance at intervals not exceeding five years, to be performed at an approved shore-based maintenance facility.

9.14   Radio personnel

Every vessel should carry personnel qualified for distress and safety radiocommunications purposes, to the satisfaction of the Competent Authority, any one of whom should be designated to have primary responsibility for radiocommunications during distress incidents. The personnel should be holders of certificates specified in the Radio Regulation as appropriate. Alternatively, national certificates based on the same requirements as the Radio Regulation, but taking account of particular local circumstances, may be issued.

9.15   Radio records

A record should be kept, to the satisfaction of the Competent Authority and as required by the Radio Regulations, of all incidents connected with the radiocommunication service which appear to be of importance to safety of life at sea.

9.16   Position-updating

All two-way communication equipment carried on board a vessel to which this chapter applies which is capable of automatically including the vessel's position in the distress alert should be automatically provided with this information from an internal or external navigation receiver, if either is installed. If such a receiver is not installed, the vessel's position and the time at which the position was determined should be manually updated at intervals not exceeding four hours, while the vessel is under way, so that it is always ready for transmission by the equipment.

Chapter X
Shipborne navigational equipment and arrangements

10.1 Shipborne navigational equipment*

10.1.1 Vessels should be fitted with:

   .1 a standard magnetic compass, except as provided in 10.1.5;

   .2 adequate means of communication between the standard compass position and the normal navigation control position, to the satisfaction of the Competent Authority; and

   .3 means for taking bearings as nearly as practicable over an arc of the horizon of 360°.

10.1.2 The magnetic compass referred to in 10.1.1 should be properly adjusted and its table or curve of residual deviations should be available at all times.

10.1.3 A spare magnetic compass, interchangeable with the standard compass, should be carried by vessels of 35 m in length and over, unless a steering compass or a gyro-compass is fitted.

10.1.4 It should be possible to read the compasses by day and by night. It should also be possible to take bearings by day or by night using the standard or steering compass or a pelorus. Magnetic compasses should be provided with means for adjustment; securing devices for compasses and compensators should be made of non-magnetic materials. Compasses should be sited as near the fore-and-aft line of the vessel as practicable, with the lubber line, as accurately as possible, parallel with the fore-and-aft line. Compasses should comply with the requirements of the Competent Authority.

10.1.5 The Competent Authority, if it considers it unreasonable or unnecessary to require a standard magnetic compass, may exempt individual vessels or classes of vessels from these requirements if the nature of the voyage, the vessel's proximity to land or the type of vessel does not warrant a standard compass, provided that a suitable steering compass is in all cases carried.

---

* Refer to the Recommendation on the carriage of electronic position-fixing equipment, adopted by the Organization by resolution A.156(ES.IV), and the World-Wide Radionavigation System, adopted by the Organization by resolution A.953(23).

10.1.6 Vessels of 45 m in length and over should be fitted with a gyro-compass complying with the following requirements:

    .1   the master gyro-compass or a gyro-repeater should be clearly readable by the helmsman at the main steering position; and

    .2   on vessels of 75 m in length and over, a gyro-repeater or gyro-repeaters should be provided and should be suitably placed for taking bearings as nearly as practicable over an arc of the horizon of 360°.

10.1.7 Vessels with emergency steering positions should at least be provided with a telephone or other means of communication for relaying heading information to such positions. In addition, vessels of 45 m in length and over equipped with gyro-compass should be provided with arrangements for supplying visual compass readings to the emergency steering position.

10.1.8 In vessels equipped with an auto-pilot system actuated by a magnetic sensor, which does not indicate the vessel's heading, suitable means should be provided to show this information. Auto-pilot systems should comply with the requirements of the Competent Authority.

10.1.9 Vessels of 35 m in length and over should be fitted with a radar installation. The radar installation should be capable of operating in the 9 GHz frequency band. Vessels of 35 m in length and over but less than 45 m may be exempted from compliance with the requirements of 10.1.16 at the discretion of the Competent Authority, provided that the equipment is fully compatible with the radar transponder for search and rescue.

10.1.10 Facilities for plotting radar readings should be provided on the navigation bridge of vessels required by 10.1.9 to be fitted with a radar installation. In vessels of 75 m in length and over, the plotting facilities should be at least as effective as a reflection plotter.

10.1.11 Vessels of 45 m in length and over should be fitted with an echo-sounding device.

10.1.12 Vessels of less than 45 m in length should be provided with suitable means, to the satisfaction of the Competent Authority, for determining the depth of water under the vessel. Where fish-finding devices are fitted they could be used for that purpose.

10.1.13 Vessels of 45 m in length and over should be fitted with a device to indicate speed and distance.

10.1.14 Vessels of 45 m in length and over should be fitted with indicators showing the rudder angle, the rate of revolution of each propeller and, in addition, if fitted with variable-pitch propellers or lateral thrust propellers, the pitch and operational mode of such propellers. All these indicators should be readable from the conning position.

10.1.15 Vessels of 75 m in length and over should be fitted with a receiver for radionavigation system, or other means, suitable for use at all times

Shipborne navigational equipment and arrangements

throughout the intended voyage to establish and update the ship's position by automatic means. The Competent Authority may exempt a vessel from this requirement if it considers it unreasonable or unnecessary for such apparatus to be carried or if the vessel is provided with other radionavigation equipment suitable for use throughout its intended voyages.

10.1.16 All equipment fitted in compliance with this section should be of a type approved at the discretion of the Competent Authority. Equipment installed on board vessels should conform to appropriate performance standards. Such standards, wherever applicable, should not be inferior to those adopted by the Organization.*

---

* Refer to the following resolutions adopted by the Organization:

Recommendation on general requirements for shipborne radio equipment forming part of the GMDSS and for electronic navigational aids (resolution A.694(17));

Recommendation on performance standards for magnetic compasses (resolution. A.382(X));

Recommendation on performance standards for gyro-compasses (resolution A.424(XI));

Recommendation on performance standards for radar equipment (resolution MSC.64(67), annex 4);

Performance standards for automatic radar plotting aids (ARPAs) (resolution A.823(19));

Recommendation on performance standards for echo-sounding equipment (resolution A.224(VII), as amended by resolution MSC.74(69), annex 4);

Recommendation on performance standards for devices to indicate speed and distance (resolution A.824(19), as amended by resolution MSC.96(72));

Performance standards for rate-of-turn indicators (resolution A.526(13));

Recommendation on unification of performance standards for navigational equipment (resolution A.575(14));

Recommendation on methods of measuring noise levels at listening posts (resolution A.343(IX));

Recommendation on performance standards for shipborne global positioning system receiver equipment (resolution A.819(19), as amended by resolution MSC.112(73));

Recommendation on performance standards for shipborne GLONASS receiver equipment (resolution MSC.53(66), as amended by resolution MSC.113(73));

Recommendation on performance standards for combined GPS/GLONASS receiver equipment (resolution MSC.74(69), annex 1, as amended by resolution MSC.115(73));

Recommendation on performance standards for heading control systems (resolution MSC.64(67), annex 3);

Recommendation on performance standards for shipborne Loran-C and Chayka receivers (resolution A.818(19));

Recommendation on performance standards for shipborne DGPS and DGLONASS maritime radio beacon receiver equipment (resolution MSC.64(67), annex 2, as amended by resolution MSC.114(73));

Recommendation on performance standards for track control systems (resolution MSC.74(69), annex 2);

Recommendation on performance standards for a universal shipborne automatic identification system (AIS) (resolution MSC.74(69), annex 3);

Recommendation on performance standards for radar reflectors (resolution A.384(X), as amended by resolution MSC.164(78));

Recommendation on performance standards for sound reception systems (resolution MSC.86(70), annex 1); and

Recommendation on performance standards for voyage data recorders (VDRs) (resolution A.861(20)).

10.2 Nautical instruments and publications

10.2.1 Suitable nautical instruments, adequate and up-to-date charts, sailing directions, lists of lights, notices to mariners, tide tables and all other nautical publications necessary for the intended voyage, to the satisfaction of the Competent Authority, should be carried on board.

.1 An electronic chart display and information system (ECDIS)* may be accepted as meeting the chart carriage requirements of this subparagraph.

.2 Back-up arrangements should be provided to meet the functional requirements of 10.2.1.1 above, if this function is partly or fully fulfilled by electronic means.†

10.3 Signalling equipment

10.3.1 Attention is drawn to the need to provide the equipment to comply in every respect with the requirements of the International Regulations for Preventing Collisions at Sea, 1972, as amended.

10.3.2 Lights, shapes and flags should be provided to indicate that the vessel is engaged in any specific operation for which such signals are used.

10.3.3 A daylight signalling lamp‡ should be provided, the operation of which is not solely dependent upon the main source of electrical power. The power supply should in any case include a portable battery.

10.3.4 Vessels intended for fishing operations in unlimited sea areas and vessels of 45 m in length and over should be provided with a full complement of flags and pendants to enable communications to be sent using the International Code of Signals.

10.3.5 All vessels which are required to carry radio installations should carry the *International Code of Signals*.

10.3.6 The *International Code of Signals* should also be carried by any other vessel which, in the opinion of the Competent Authority, has a need to use it. Nevertheless, such vessels should carry at least the table of life-saving signals contained in the *International Code of Signals*.

---

* Refer to Performance standards for electronic chart display and information systems (ECDIS) (resolution A.817(19), as amended by resolutions MSC.64(67), annex 5, and MSC.86(70), annex 4, as appropriate).

† An appropriate folio of paper nautical charts may be used as a back-up arrangement for ECDIS. Other back-up arrangements for ECDIS are acceptable (see appendix 6 to the annex to resolution A.817(19), as amended).

‡ Recommendation on performance standards for daylight signalling lamps (resolution MSC.95(72)).

10.4 Navigating bridge visibility

10.4.1 Vessels of 45 m in length and over should meet the following requirements:

.1 the view of the sea surface from the conning position should not be obscured by more than two vessel lengths, or 500 m, whichever is less, forward of the bow to 10° on either side irrespective of the vessel's draught and trim;

.2 no blind sector, caused by fishing gear or other obstructions outside of the wheelhouse forward of the beam which obstructs the view of the sea surface as seen from the conning position, should exceed 10°. The total arc of blind sectors should not exceed 20°. The clear sectors between blind sectors should be at least 5°. However, in the view described in .1 above, each individual blind sector should not exceed 5°;

.3 the height of the lower edge of the navigation bridge front windows above the bridge deck should be kept as low as possible. In no case should the lower edge present an obstruction to the forward view as described in this paragraph;

.4 the upper edge of the navigation bridge front windows should allow a forward view of the horizon for a person with a height of eye of 1800 mm above the bridge deck at the conning position when the vessel is pitching in heavy seas. However, the Competent Authority, being satisfied that a 1800 mm height of eye is unreasonable and impractical, may reduce the height of eye, but not to less than 1600 mm;

.5 the horizontal field of vision from the conning position should extend over an arc of not less than 225°, that is from right ahead to not less than 22.5° abaft the beam on either side of the vessel;

.6 from each bridge wing the horizontal field of vision should extend over an arc of at least 225°, that is from at least 45° on the opposite bow through right ahead and then from right ahead to right astern through 180° on the same side of the vessel;

.7 from the main steering position the horizontal field of vision should extend over an arc from right ahead to at least 60° on each side of the vessel;

.8 the vessel's side should be visible from the bridge wing; and

.9 windows should meet the following requirements:

.9.1 framing between navigation bridge windows should be kept to a minimum and not be installed immediately forward of any workstation;

143

- **.9.2** to help avoid reflections, the bridge front windows should be inclined from the vertical plane top out, at an angle of not less than 10° and not more than 25°;
- **.9.3** polarized and tinted windows should not be fitted; and
- **.9.4** a clear view through at least two of the navigation bridge front windows and, depending on the bridge configuration, an additional number of clear view windows should be provided at all times regardless of weather conditions.

10.4.2 Existing vessels should, where practicable, meet the requirements of 10.4.1.1 and 10.4.1.2. However, structural alterations or additional equipment need not be required.

10.4.3 On vessels of unconventional design which, in the opinion of the Competent Authority, cannot meet the requirements of 10.4, arrangements should be provided to achieve a level of visibility that is as near as practicable to that stated in 10.4.1.

10.4.4 For vessels below 45 m in length, the Competent Authority should determine which of the requirements contained in 10.4.1 to 10.4.3 should apply, wholly or in part.

10.5   Pilot transfer arrangements

10.5.1 Vessels engaged on voyages in the course of which pilots are likely to be employed should be provided with pilot transfer arrangements.

10.5.2 When a vessel is at sea, similar arrangements to 10.5.1 should be provided for fisheries inspectors.

10.5.3 Such transfer arrangements should comply with the provisions of annex VI in this part of the Code.

10.6   Documents

Vessels should be supplied with appropriate logs, certificates and other documents in accordance with the provisions of international and national regulations.

Chapter XI
Crew accommodation

11.1 General

11.1.1 Before the construction of a fishing vessel, and before the crew accommodation of an existing fishing vessel is substantially altered or reconstructed, detailed plans of, and information concerning, the accommodation should be submitted to the Competent Authority, or an entity authorized by the Competent Authority, for approval.

11.1.2 In vessels intended for fishing on the high seas or for distant-water fishing in waters of States other than those of the flag State, and carrying a crew of more than 20, consideration should be given to the provision of separate messroom accommodation for the skipper and officers and, where applicable, observers and scientists.

11.1.3 Location, structure and arrangement of crew accommodation spaces and means of access thereto should be such as to ensure adequate security, protection against weather and sea and insulation from heat and cold, undue noise, vibration or effluvia from other spaces. In particular, the insulation material to be applied to bulkheads and deckheads of machinery spaces adjacent to crew accommodation should be of a type approved by the Competent Authority.

11.1.4 Where, in view of operational requirements, the Competent Authority has permitted sleeping rooms to be placed in the fore part of the vessel, they should be placed aft of the collision bulkhead and, to the extent possible, not below the working deck.

11.1.5 Bulkheads and decks between accommodation spaces and fish-holds; machinery spaces; fuel tanks; galleys, engine, deck and other storerooms; drying rooms, communal wash-places or water-closets, should be so constructed as to prevent the infiltration of fumes and odours. Direct openings into sleeping rooms from such places should be avoided wherever reasonable or practicable. That part of bulkheads separating such places from sleeping rooms, and also external bulkheads, should be gastight and, where necessary, should prevent the passage of water.

11.1.6 Where corridors are provided in crew accommodation, these should be as wide as possible but should not be less than 700 mm and be fitted with handrails on at least one side. Where doors open outwards into a

Chapter XI

passageway, there should be sufficient space to pass the door when it is open at a right angle to the passageway.

11.1.7 Accommodation spaces should be adequately insulated to prevent loss of heat, condensation or overheating. Care should be taken to provide protection from heat effects of steam and/or hot-water service pipes.

11.1.8 Fuel oil, sounding and hydraulic oil pipes, high-voltage electrical wiring for winch machinery or steam piping, except steam heating systems, should not be led through accommodation spaces unless such arrangement is approved by the Competent Authority.

11.1.9 In the choice of materials used for construction of accommodation spaces, account should be taken of properties potentially harmful to the health of personnel, or likely to harbour vermin and mould. Surfaces, including decks, of accommodation and furnishings should be of a kind easily kept clean and hygienic, as well as impervious to damp. Bulkhead and deckhead surfaces, if painted, should be light in colour and the paint specification should be to the approval of the Competent Authority. Other surface coverings, such as lime wash, should not be used.

11.1.10 Where the deck covering is of composition material, the connection to the side of the vessel, bulkheads and partitions should be rounded to avoid crevices.

11.1.11 All practical measures should be taken to protect crew accommodation and furnishings against the admission of insects and other pests.

11.1.12 Overhead exposed decks over crew accommodation should be sheathed with wood or equivalent insulation.

11.1.13 The electrical switchboard should be so arranged that when the shore power connection is made, power would be available for crew accommodation lighting, ventilation systems and, where applicable, heating and cooking facilities.

11.1.14 Access to ordinary exits and emergency exits should be marked with direction indicators. Exits should be marked in a conspicuous manner above or beside the door.

11.2 Lighting, heating and ventilation

11.2.1 All crew accommodation spaces should be adequately lit, as far as possible, by natural lighting. Such spaces should also be equipped with adequate artificial light. Artificial lighting should be in accordance with accepted standards of visual comfort in living spaces.

11.2.2 The minimum standards for natural lighting in crew accommodation should be such as to permit a person with normal vision to read an ordinary newspaper on a clear day.

Crew accommodation

11.2.3 If there are not two independent sources of electricity for lighting, additional lighting should be provided by properly constructed lamps or lighting apparatus for emergency use.

11.2.4 An adequate reading light should be provided for every berth in addition to the normal lighting of the cabin.

11.2.5 A permanent night light should, in addition to the normal lighting, be provided in sleeping rooms during the night. Messrooms and alleyways, that contain emergency escapes facilities from the crew accommodation, should also be provided with a permanent night light during the night.

11.2.6 Methods of lighting should not endanger the health or safety of the crew or the safety of the vessel.

11.2.7 Adequate heating facilities in accommodation spaces should be provided as required by climatic conditions. Heating facilities should be capable of maintaining a satisfactory air temperature in crew accommodation under normal conditions of service of a fishing vessel and as prescribed by the Competent Authority. The accommodation should be capable of being heated sufficiently to maintain a minimum temperature of $+22°C$ in all dayrooms at an outside temperature of $-15°C$.

11.2.8 Facilities for heating should be designed so as not to endanger the health or safety of the crew or the safety of the vessel.

11.2.9 Heating by means of open fires should be prohibited.

11.2.9.1 Accommodation spaces should be adequately ventilated at all times when the crew is expected to remain on board. Ventilation systems should be capable of control so as to maintain the air in a satisfactory condition and to ensure a sufficiency of air movement in all conditions of weather and climate. The ventilation of galley, sanitary and hospital spaces should be to the open air and, unless fitted with a mechanical ventilation system approved by the Competent Authority, be independent from that for other crew accommodation.

11.2.9.2 Accommodation spaces of vessels regularly engaged on voyages in the tropics and in similar climatic conditions, except in deckhouses with satisfactory natural ventilation, should be equipped with mechanical ventilation and, if necessary, with additional electric fans or air conditioning, in particular in mess rooms. When necessary to ensure satisfactory ventilation, vessels engaged elsewhere should be equipped either with mechanical means of ventilation or with electric fans.

11.2.10 Vessels fitted with air conditioning should carry a suitable gas detector

11.2.11 Drying rooms or lockers for working clothes and oilskin lockers should have adequate ventilation that is independent of other spaces. The exhaust from such spaces should be well clear of the air intakes of the ventilation systems for other spaces.

11.3 Sleeping rooms

11.3.1 Sleeping rooms should be so planned and equipped as to ensure reasonable comfort for the occupants and to facilitate tidiness. The clear headroom should, whenever possible, be not less than 2 m.

11.3.2 The floor area per person of sleeping rooms, excluding space occupied by berths and lockers, should not be less than:

.1   1 m^2 in vessels of 24 m but below 45 m in length; and

.2   1.5 m^2 in vessels of 45 m in length or over.

11.3.3 Wherever reasonable and practicable with respect to the size, type or intended service of a vessel, the number of persons allowed to occupy each sleeping room should not be more than four persons in vessels of 37 m in length and over and six persons in vessels of less than 37 m in length. Sleeping rooms for officers should be for one person wherever possible and in no case should the sleeping room contain more than two berths.

11.3.4 The maximum number of persons to be accommodated in any sleeping room should be clearly and indelibly marked in the room where it could be conveniently seen. Where appropriate, a notice should also be posted in a language understood by the majority of the occupants.

11.3.5 Each member of the crew should be provided with an individual berth, the minimum inside dimensions of which should, wherever practicable, be 1.9 m by 680 mm.

11.3.6 Berths should not be placed side by side in such a way that access to one berth can be obtained only over another. Berths should not be arranged in tiers of more than two. The lower berth in a double tier should not be less than 300 mm above the deck; the upper berth should be placed approximately midway between the bottom of the lower berth and the lower side of the deck head beams.

11.3.7 Where the upper berth in a tier overlaps a lower berth, the underside of the upper berth should be fitted with a dustproof bottom of wood, canvas or other material.

11.3.8 The framework and the lee-board, if any, of a berth should be of approved material, hard, smooth and not likely to corrode or to harbour vermin.

11.3.9 If tubular frames are used for the construction of berths, they should be completely sealed and without perforations that would give access to vermin.

11.3.10 Suitable bedding should be provided for the crew. Mattresses should not be of a type that is liable to develop toxic fumes in case of fire or to harbour vermin. Mattresses should be provided with a cover of fire-retardant material.

11.3.11 Wherever reasonable and practicable with respect to the size, type or intended service of a vessel, the furnishings of sleeping rooms should include both a fitted cupboard, preferably with an integral lock, and a drawer for each occupant. Sleeping rooms should also be fitted with a satisfactory table or desk, adequate and proper seating, curtains for sidelights, a mirror, cabinets for toilet requisites, a bookshelf and coat hooks.

11.3.12 Where fishers are carried on board for the sole purpose of operating from small boats carried by the fishing vessel and are not part of the crew, suitable sleeping accommodation, sanitary and messroom facilities should be provided. Due to the different nature and frequency of their operations to those of the crew, such facilities should preferably be separate to those provided for the crew normally assigned watchkeeping duties.

11.4 Messrooms

11.4.1 Messroom accommodation separate from sleeping quarters should be provided in all vessels regularly carrying a crew of more than ten persons. Wherever reasonable and practicable, it should be provided also in vessels carrying a smaller crew.

11.4.2 Messrooms should be as close as practicable to the galley.

11.4.3 The dimensions and equipment of each messroom should be sufficient for the number of persons likely to use it at any one time.

11.4.4 The furnishings of messrooms should include tables and approved seats sufficient for the number of persons likely to use them at any one time. The tops of tables and seats should be free of sharp edges and be of damp-resisting material without cracks and easily kept clean.

11.4.5 Where pantries are not accessible from messrooms, adequate lockers for mess utensils and proper facilities for washing should be provided.

11.4.6 Messrooms should be planned, furnished and equipped to provide appropriate facilities for recreation.

11.4.7 Whenever possible, a separate recreation room should be provided for the crew.

11.5 Sanitary facilities

11.5.1 Sufficient sanitary facilities including washbasins and tubs and/or shower-baths and water closets should be provided on a scale approved by the Competent Authority. Wherever practicable, such facilities should be provided as follows:

    .1   one tub and/or shower-bath for every eight persons;

    .2   one water closet for every eight persons or less; and

Chapter XI

.3 one washbasin for every six persons or less.

Provided that when the number of persons exceeds an even multiple of the specified number by less than one half of the specified number, this surplus may be ignored for the purpose of this paragraph.

11.5.2 Where there is more than one water closet in a compartment, they should be adequately screened to ensure privacy.

11.5.3 In general, water closets should be situated convenient to, but separate from, sleeping rooms, messrooms and washrooms.

11.5.4 In cases where a water closet is provided with direct access from sleeping places that are intended for not more than two persons, the access should be so constructed as to provide a reasonable seal when closed. Such water closets may also house washing facilities and should be provided with a separate means of ventilation and should not ventilate to or through the adjacent sleeping space.

11.5.5 Cold fresh water and hot fresh water or means of heating fresh water should be available in all washplaces.

11.5.6 The deck area of washplaces should have a covering of durable material, easily cleaned, impervious to damp and properly drained. The deck covering should be carried up the sides of the compartment to a height of not less than 0.2 m and be adequately sealed at all joints to prevent the ingress of water and damp.

11.5.7 The bulkheads should be of steel or other approved material and should be watertight to a height of at least 0.25 m above the deck to allow for effective sealing of the deck covering.

11.5.8 All sanitary equipment and systems should be of a design, construction and size approved by the Competent Authority. In particular, showers should have anti-scalding valves of an approved type, sufficient drainage should be provided, and soil and waste discharge pipes should be of adequate dimensions and constructed so as to facilitate cleaning. International standards concerning shipboard sanitary facilities contained in the WHO *Guide to Ship Sanitation*, 1967, as amended, may serve as guidance.

11.5.9 Soil and waste discharge pipes should not pass through fresh water or drinking water tanks or, where practicable, provision stores. Neither should they, where practicable, pass overhead in messrooms or sleeping accommodation. Such pipes should be fitted with anti-siphon closures.

11.5.10 Facilities for washing and drying clothes should be provided on a scale appropriate to the number of the crew and the duration of intended voyages. These facilities should include an adequate supply of cold fresh water and hot fresh water or means of heating fresh water. Wherever reasonable and practicable, separate laundry accommodation should be provided.

11.6 Potable water facilities

Filling, storage and distribution for potable water should be designed to preclude any possibility of water contamination or overheating. In this connection, the relevant international standards laid down in the WHO *Guide to Ship Sanitation*, 1967, as amended, should be followed.

11.7 Provision stores

Provision store-rooms of adequate capacity should be provided which can be kept cool, dry and well ventilated in order to avoid deterioration of the stores. Where necessary, taking into consideration the area of operation and the duration of the voyage, refrigerators or other low-temperature storage space should be provided. It should be possible to keep a temperature in refrigerating rooms or similar rooms of between $-1°C$ and $+4°C$ in all climatic conditions. Vessels whose area of operation requires foodstuffs to be frozen during storage should be fitted with chest freezers, upright freezers or freezing rooms. It should be possible to keep a temperature of $-25°C$ or lower in all climatic conditions and store fish separate from other foodstuffs. The temperature in refrigerating and freezing rooms should be capable of being read from the outside. Doors to refrigerating and freezing rooms should be capable of being opened from either side. An alarm system should be arranged from the refrigerating and freezing room to the galley or other appropriate location if such rooms are large enough for personnel to enter them.

11.8 Cooking and beverage facilities

11.8.1 Satisfactory cooking appliances and equipment should be provided and should, wherever practicable, be fitted in a separate galley.

11.8.2 Cooking appliances should be fitted with fail-safe devices in the event of failure of the power source or fuel. Supplies of fuel in the form of gas or oil should not be stored in the galley.

11.8.3 Galleys should be of adequate dimensions for the purpose and have sufficient storage space and satisfactory drainage. International standards concerning shipboard food sanitation laid down in the WHO *Guide to Ship Sanitation*, 1967, as amended, may serve as guidance.

11.8.4 The galley should be equipped with cooking utensils, the necessary number of cupboards, shelves, sinks and dish racks of rustproof material and with satisfactory drainage. Drinking water should be supplied to the galley by means of pipes. Where it is supplied under pressure, the system should be protected against backflow. Where hot water is not supplied to the galley, a water heater should be fitted.

11.8.5 The galley should be fitted with suitable facilities for the preparation of hot drinks for the crew at all times.

11.8.6 A domestic refrigerator of sufficient capacity for the number of persons using each messroom should be provided. Facilities should also be provided for hot beverages and cool water for the benefit of the crew.

11.9 Hospital accommodation

A sick bay or equivalent together with suitable sanitary facilities should be provided in vessels of 45 m in length and over intended for fishing operations in sea areas beyond 50 nautical miles from a place of shelter. In vessels of 100 m in length or over, the sick bay should be equipped with a detachable and portable swivel bed and should be designed to facilitate removal of a patient in the portable bed.

Annex I

Illustration of terms used in the definitions

LENGTH (L)

1. L = 0.96 OF THE TOTAL LENGTH ON A WATERLINE AT 85 PER CENT OF LEAST DEPTH

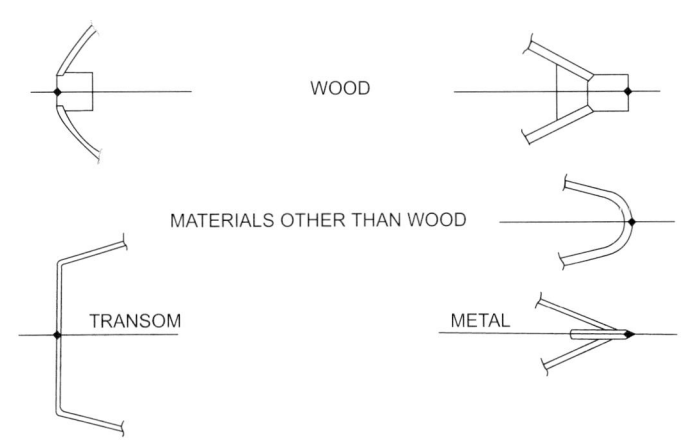

2. L = LENGTH ON A WATERLINE AT 85 PER CENT OF LEAST DEPTH BETWEEN THE STEM AND THE AXIS OF THE RUDDER STOCK

Figure 1

Annex I

BREADTH (B)

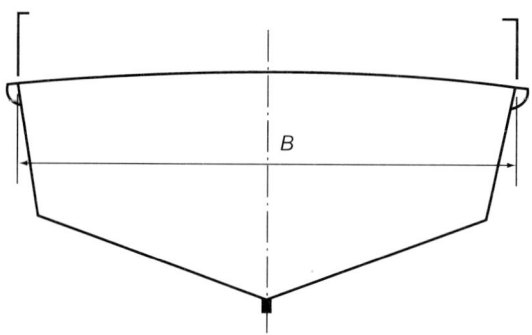

LEAST DEPTH

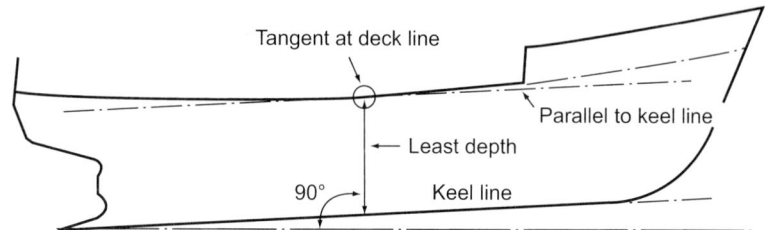

Figure 2

Illustration of terms used in the definitions

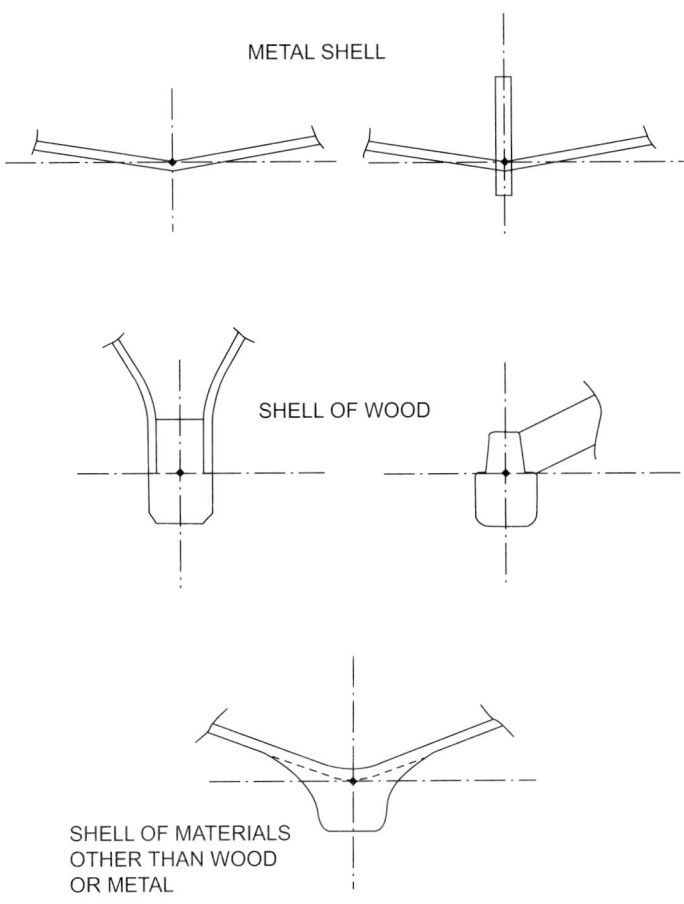

Figure 3

Annex I

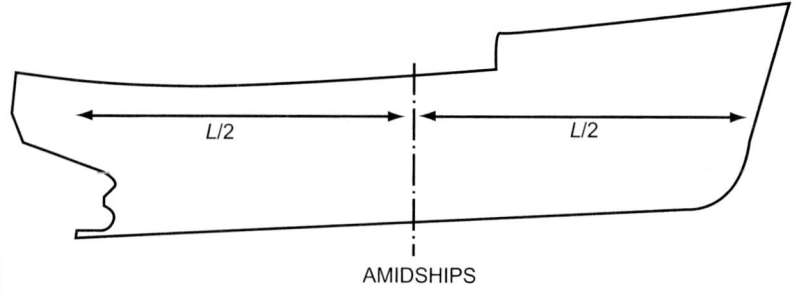

Figure 4

Annex II

Recommended practice for anchor and mooring equipment

1   The characteristics of anchors, chain, wires, towlines and mooring lines should be determined in accordance with the attached table, based on an equipment number "EN" as follows.

$$EN = \Delta^{\frac{2}{3}} + 2B(a + \Sigma h_j) + 0.1A$$

where:

- Δ moulded displacement, in tonnes, to the maximum design waterline.
- B breadth, in metres, as defined in 1.2.5.
- a distance, in metres, from the maximum design waterline to the upper edge of the uppermost complete deck at side amidships.
- h_j height, in metres, on the centreline of each tier of deckhouses having a breadth greater than $B/4$. For the lowest tier, h_j should be measured at centreline from the upper deck or from a notional deck line where there is a local discontinuity in the upper deck. When calculating h_j, sheer and trim should be ignored.
- A area, in square metres, in profile view of the hull, within L as defined in 1.2.20 and of superstructures and deckhouses above the maximum design waterline having a width greater than $B/4$. Screens and bulwarks more than 1.5 m in height should be regarded as parts of deckhouses when determining h_j and A.

Anchors and chains

2   Vessels should be fitted with at least two anchors which should be located at the bow.

3   The weight of each anchor should be in accordance with the table given in this annex.

4   "High holding power anchors" of a design approved by the Competent Authority may be used as bower anchors; the weight of each such anchor may be 75% of the table weight given in this annex.

Annex II

5   The Competent Authority may require increased anchor equipment for vessels fishing in very rough waters and/or may permit reduction in the equipment for vessels operating in sheltered waters.

6   Anchors with a weight of 150 kg and above should be fitted in hawse pipes, skids or a similar arrangement that is suitable for the quick and safe operation in dropping and hoisting the anchors. If the weight of each of the anchors is below 300 kg, but greater than 150 kg, it may be accepted that only one of the anchors need be fitted in a hawse pipe or skid. Anchors should also be secured in the stowed position by means of a locking or lashing device.

7   In general, anchors should be fitted with anchor chains. The length and dimension of each anchor chain should be in accordance with the table given in this annex.

8   For vessels less than 45 m in length, the chain of one anchor may be replaced with anchor wires of equal strength provided a chain meeting the requirements given in the table to this annex is maintained for the second one.

9   Where anchor wires are used as a substitute for anchor chains, their length should be equal to 1.5 times the corresponding tabular length of chain. In addition, a chain of not less than 12.5 m in length and of the same specifications as set out in the table to this annex should be provided between anchor and anchor wire.

10   Where the Competent Authority has authorized the use of trawl warp as anchor wire, it should be satisfied that the arrangement does not reduce the efficiency required for the quick and safe operation in dropping and hoisting the anchors and for holding the vessel at anchor in all foreseeable service conditions. The requirements for a trawl warp should not be less than that required for anchor wire.

Anchor handling

11   Fishing vessels provided with anchors of 150 kg or above should be fitted with a windlass. The windlass should be fitted with a messenger wheel and/or drum for each anchor and means for the release of each messenger wheel or drum.

12   It should not be possible to carry the chains forward to the hawse pipe, skid or similar arrangement without the chain passing over the messenger wheels. When anchor wire is used, it should pass over a roller adjacent to the hawse pipe to avoid chafing.

13   The windlass, its support and its brakes should be capable of absorbing a static tension of at least 45% of the breaking strength of the anchor chain or anchor wire without the occurrence of any lasting deformations and without the brake losing its hold. Furthermore, a chain stopper or wire nipper should be fitted between the windlass and the hawse pipe or similar for each anchor chain or anchor wire capable of holding the vessel while at anchor. If chain stoppers or wire nippers are not fitted, the windlass, its support and its brake should be capable of absorbing a static tension of at least 80% of the breaking strength of the anchor chain or anchor wire. The chain stopper or wire nipper and their supports should be capable of absorbing a static tension of at least 80% of the breaking strength of the anchor chain/wire without the occurrence of any lasting deformations and without the chain stopper or wire nipper losing its hold.

14 If the trawl winch is fitted with messenger wheels, etc. and meets the requirements set out in 11, 12 and 13, such a winch may be used as a windlass.

15 Fishing vessels which have been authorized to use trawl warp as anchor wire may use their trawl winch as a windlass provided the trawl warp can be wound on a drum with a braking device that is independent of the actual trawl warps in use for fishing. Lead blocks and guide rollers should be suitably fitted and arranged to prevent the warps from chafing at the deckhouses, superstructures, deck plating and equipment on deck.

16 If a vessel has lost its anchors, and it is not immediately possible to re-acquire them, the Competent Authority, after having assessed the conditions applying to the vessel, as given in 5, may permit otterboards/ trawl doors with a least the same weight for anchors given in the table to this annex to be used for a limited period of time.

Towing lines

17 Vessels should be provided with at least one tow line with a length and breaking strength in accordance with the table given in this annex. It should be appropriately located so that it is possible to make it ready for use at sea. The tow line may be replaced by one of the fishing vessel's trawl warps if this has at least a similar length and breaking strength. If warp is used, a length of rope of at least 12.5 m with a minimum breaking strength as given in the table for the tow line should also be provided and attached to the warp.

Mooring equipment

18 Vessels should be provided with suitable cleats and bollards as well as hawseholes in order to moor the vessel securely. The number of bollards, etc. should be determined in each individual case dependent on the size and deck arrangement of the vessel. The number should be sufficient to make it possible to fasten both the mooring line and a spring on each bollard on each side forward and aft. At least three bollards should be fitted forward, and at least two abaft of amidships. Cleats and bollards should be of such a size that it is possible to accommodate at least four turns of the mooring lines or tow line below the horns of the cleat or the

Annex II

upper protruding edge of the bollard. The area where cleats and bollards are to be fastened should be securely reinforced.

19 The vessel should be provided with at least four mooring lines, each of a length and breaking strength in accordance with the table given in this annex.

Table

Equipment number		Stockless bower anchors		Stud link chain cables for bower anchors				Towline		Mooring lines	
Exceeding	Not exceeding	Number	Weight per anchor (kg)	Total length (m)	Diameter (mm) Mild steel	Diameter (mm) Special quality steel		Minimum length of each line (m)	Minimum breaking strength (kN)	Minimum length of each line (m)	Minimum breaking strength (kN)
50	60	2	120	192.5	12.5	–		180	98	60	34
60	70	2	140	192.5	12.5	–		180	98	80	34
70	80	2	160	220	14	12.5		180	98	100	37
80	90	2	180	220	14	12.5		180	98	100	37
90	100	2	210	220	16	14		180	98	110	39
100	110	2	240	220	16	14		180	98	110	39
110	120	2	270	247.5	17.5	16		180	98	110	44
120	130	2	300	247.5	17.5	16		180	98	110	44
130	140	2	340	275	19	17.5		180	98	120	49
140	150	2	390	275	19	17.5		180	98	120	49
150	175	2	480	275	22	19		180	98	120	54
175	205	2	570	302.5	24	20.5		180	112	120	59
205	240	2	660	302.5	26	22		180	129	120	64
240	280	2	780	330	28	24		180	150	120	69
280	320	2	900	357.5	30	26		180	174	140	74
320	360	2	1020	357.5	32	28		180	207	140	78
360	400	2	1140	385	34	30		180	224	140	88
400	450	2	1290	385	36	32		180	250	140	98
450	500	2	1440	412.5	38	34		180	277	140	108
500	550	2	1590	412.5	40	34		190	306	160	123
550	600	2	1740	440	42	36		190	338	160	128
600	660	2	1920	440	44	38		190	371	160	132
660	720	2	2100	440	46	40		190	406	160	137

Annex III

Recommended practice on portable fish-hold divisions*

1   Recognizing the desirability of ensuring the adequate strength of scantlings of portable fish-hold divisions*, studies on national practices have been carried out, resulting in the establishment of certain formulae for scantlings, which are recommended to Administrations for their guidance.

2   These formulae represent the average of a wide range of experience covering all types of vessels operating in all sea areas, and in conditions likely to impose the maximum loading on a division. Alternative scantlings might, however, be accepted where experience has shown that these are more appropriate.

3   According to the basic type of construction, the following formulae are recommended for vertical fish-hold divisions:

   .1 Vertical steel uprights and horizontal wooden boards

   Minimum section modulus of vertical steel uprights

   $$Z = 4\rho sbh^2 \tag{1}$$

   Minimum thickness of horizontal wooden boards

   $$t = \sqrt{8\rho sb^2} \tag{2}$$

   .2 Horizontal steel beams and vertical wooden boards

   Minimum section modulus of horizontal steel beams

   $$Z = 4\rho sHS^2 \tag{3}$$

   Minimum thickness of vertical wooden boards

   $$t = \sqrt{3.6\rho sh^2} \tag{4}$$

where:

Z = section modulus, in cm³.
t = thickness of wooden board, in cm.
ρ = density of cargo, in t/m³.
s = maximum transverse distance between any two adjacent longitudinal divisions or line of supports, in m.

---

* Appendix V of the annex to Assembly resolution A.168(ES.IV) incorporating subparagraphs 4(g) and 4(h) adopted by the eighth Assembly.

h = maximum vertical span of a column taken to be the hold depth, in m.

b = maximum longitudinal distance between any two adjacent transverse divisions or line of supports, in m.

H = vertical span of a division which is supported by a horizontal beam, in m.

S = horizontal distance between adjacent points of support of a horizontal beam, in m.

4 In applying the above formulae, the following notes should be observed:

.1 The formulae are applicable to longitudinal divisions. Where the divisions are athwartships, the formulae should be modified by interchanging s and b.

.2 The formulae were derived on the assumption that the loads were on one side only of the divisions. When it is known that the divisions will always be loaded on both sides, reduced scantlings may be accepted.

.3 If vertical steel uprights are permanent and well connected at both ends with the structure of the ship, reduced scantlings may be accepted depending upon the degree of security provided by the end connections.

.4 In the formula for vertical wooden boards, the full depth of the hold is assumed as the unsupported span; where the span is less, the thickness may be calculated using the reduced span.

.5 The timber used should be of sound durable quality, of a type and grade which has proved satisfactory for fish-hold divisions, and the actual finished thicknesses of boards should be those derived from the formulae. The thickness of boards made from good quality hardwood may be reduced by 12.5%.

.6 Divisions made of other materials should have strength and stiffness equivalent to those associated with the scantlings recommended for wood and steel, having regard to the comparative mechanical properties of the materials.

.7 Channelways in stanchions to take pound boards should have a depth of not less than 4 cm and the width should be equal to the pound board thickness plus 0.5 cm.

.8 Each pound board should have a length not less than the distance between the bottom of the respective channelways into which it will engage minus 1 cm.

If pound boards have shaped ends to allow a rotational manoeuvre for easy housing, the extent of end shaping should not be more than allowed by a radius equal to one half the length of the board with its centre at the mid length and depth of the board.

Annex III

5   Figures 1 and 2 illustrate the application of the formulae.

Horizontal wood boards – steel uprights

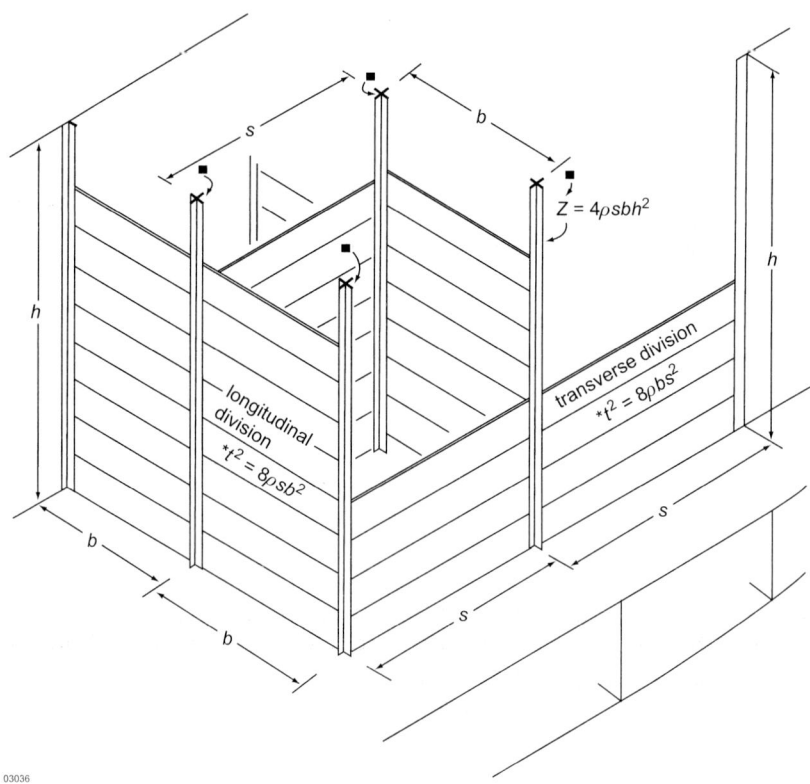

Figure 1

*Note: When the longitudinal and transverse divisional boards are interchangeable, b will equal s and the thickness by either formula will be the same. If the boards are required to be of equal thickness but varying span, the greater thickness should be used for all the boards when the section modulus is kept constant for all the uprights.

Vertical wood boards – steel beams

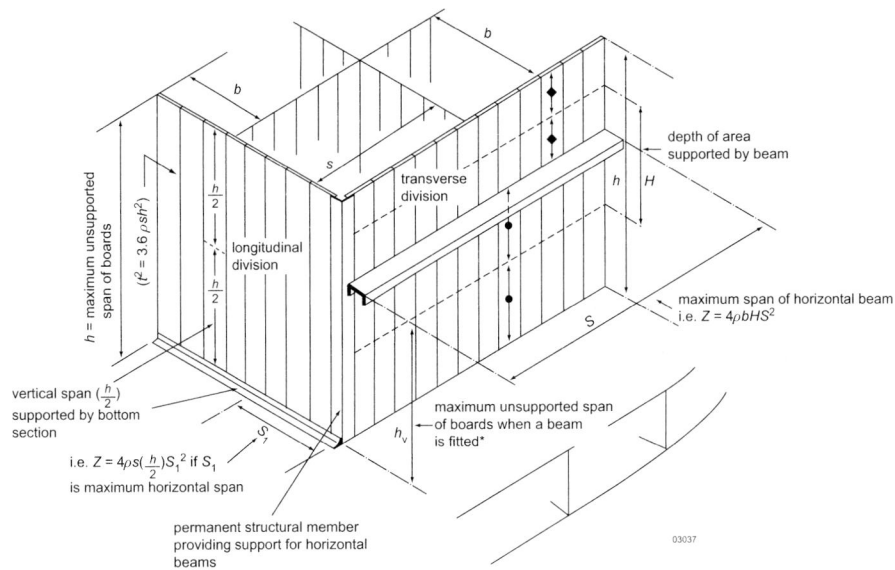

Figure 2

*Note: If no beam was fitted, the thickness of the vertical wood planks would be given by $t^2 = 3.6\rho bh^2$. The beam reduces the maximum span to h_v and the thickness is now given by $t_1^2 = 3.6\rho bh_v^2$ or $t_1 = t\left(\frac{h_v}{h}\right)$

Annex IV

Recommended practice for ammonia refrigeration systems in manned spaces

General

1   All electrical equipment on, or adjacent to, the ammonia machinery flat should be explosion-proof or of an intrinsically safe type, to the satisfaction of the Competent Authority.

2   Flame-producing devices and hot surfaces above 427°C in the machinery space should be located as remotely as practical from the ammonia machinery flat.

3   Ammonia equipment should be surrounded by an efficient water curtain and in addition, water sprays should be directed at all potential leak sources, e.g. pipe connections and flanges, compressors, etc. The water curtain and sprays should be provided with an adequate supply of water, which should be maintained under constant pressure.

4   A large capacity ventilation system, including mechanical exhaust, should be provided for the ammonia machinery flat. The system should not exhaust to another space and should be well clear of ventilator intakes to other spaces. The mechanical exhaust ventilation fan motor should be either fitted exterior to the ammonia flat or should be of an intrinsically safe type, to the satisfaction of the Competent Authority.

5   Coamings should be provided around the ammonia machinery flat.

6   Personal safety equipment, including suitable gas masks and protective clothing, should be provided inside and outside the machinery space.

7   Remote controls located in the wheelhouse or other suitable place should be provided for the following services:

   .1   the water curtain spray system;

   .2   the ammonia machinery flat ventilation system; and

   .3   the main engine.

8 Means should be provided for stopping the ammonia compressor prime movers from the wheelhouse or another suitable place.

9 Means of escape, direct to deck, from the ammonia machinery flat should be provided in addition to any other escape which may be required by the Competent Authority.

10 Drainage should be provided from machinery spaces and/or flats leading to a place where water, which could be contaminated with refrigerant, presents no danger to the vessel or to persons on board.

11 Information concerning hazards, precautions and first aid should be clearly displayed at the access to the ammonia machinery space.

Piping systems

12 Joints in steel piping systems should be butt-welded, wherever practical, to reduce the possibility of leaks. Flanged joints should be limited to connections with compressors, vessels, valves, branches for future extensions or where required for maintenance. The number of joints, whether flanged or welded, should be kept to a minimum.

13 If, for operational reasons, flexible hoses are required, the Competent Authority should be satisfied that they are suitable for ammonia service. They should be adequately protected against mechanical damage, torsion and stress.

14 To the extent possible, flexible bellows should be avoided. Where flexible bellows are proposed, the Competent Authority should be satisfied that they are only used within the recommendations of the manufacturer and adequate precautions are taken to avoid excessive vibration, mechanical damage, torsion and stress.

15 All refrigerant piping should be adequately supported and the supports or hangers should be designed to carry the weight of the pipe including contents and, where required, insulation.

16 There should be sufficient clearance around pipelines to allow for any necessary attention to flanges, screwed joints and fittings.

17 Ammonia piping should not be located in lift wells, accommodation spaces, in stairways or at entrances/exits. Pipework should also be arranged so as not to obstruct access ways and inhibit access to the machinery.

18 Special attention should be paid to the clearance around pipes passing through fire-resistant bulkheads and deckheads, which should be adequately sealed to maintain the integrity of the bulkhead or deckhead. Pipe ducts and shafts should be isolated from other spaces to resist the spread of fire.

Annex IV

Decommissioning

19 When a refrigeration system is to be decommissioned or taken out of service and dismantled, the procedure should ensure that:

.1 hazards to the personnel carrying out the process are minimized;

.2 refrigerant and oil are correctly removed for reclamation or disposal; and

.3 the system as left does not present any future hazard to personnel or to the environment due to residual content.

Annex V

Recommendations for testing lifejackets and lifebuoys*

Part 1
Prototype test for life-saving appliances

1 Lifebuoys

1.1 Temperature test

The lifebuoys should be alternately subjected to surrounding temperatures of −30°C and +65°C. These alternating cycles need not follow immediately after each other and the following procedure, repeated for a total of 10 cycles, is acceptable:

  .1 an 8 h cycle at +65°C to be completed in one day;

  .2 the specimens removed from the warm chamber that same day and left exposed under ordinary room conditions until the next day;

  .3 an 8 h cycle at −30°C to be completed the next day; and

  .4 the specimens removed from the cold chamber that same day and left exposed under ordinary room conditions until the next day.

1.2 Test for oil resistance

One of the lifebuoys should be immersed horizontally for a period of 24 h under a 100 mm head of diesel oil at normal room temperature. After this test the lifebuoy should show no sign of damage such as shrinking, cracking, swelling, dissolution or change of mechanical qualities.

---

* Refer to Standardized life-saving appliance evaluation and test report forms (MSC/Circ.980).

1.3 Fire test

The other lifebuoy should be subjected to a fire test. A test pan 30 cm × 35 cm × 6 cm should be placed in an essentially draught-free area. Water should be put in the bottom of the test pan to a depth of 1 cm followed by enough petrol to make a minimum total depth of 4 cm. The petrol should then be ignited and allowed to burn freely for 30 s. The lifebuoy should then be moved through flames in an upright, forward, free-hanging position, with the bottom of the lifebuoy 25 cm above the top edge of the test pan so that the duration of exposure to the flames is 2 s. The lifebuoys should not sustain burning or continue melting after being removed from the flames.

2 Lifejackets

2.1 Temperature cycling test

A lifejacket should be subjected to the temperature cycling as prescribed in 1.1 and should then be externally examined. If the buoyancy material has not been subjected to the tests prescribed in 2.7, the lifejacket should also be examined internally. The lifejacket materials should show no sign of damage such as shrinking, cracking, swelling, dissolution or change of mechanical qualities.

2.2 Buoyancy test

The buoyancy of the lifejacket should be measured before and after 24 h complete submersion to just below the surface in fresh water. The difference between the initial buoyancy and the final buoyancy should not exceed 5% of the initial buoyancy.

2.3 Fire test

A lifejacket should be subjected to the fire test prescribed in 1.3. The lifejacket should not sustain burning or continue melting after being removed from the flames.

2.4 Test for oil resistance

2.4.1 The lifejacket should be tested for oil resistance as prescribed in 1.2.

2.4.2 If the buoyancy material has not been subjected to the tests prescribed in 2.7, the lifejacket should also be examined internally and the effect determined. The material should show no sign of damage such as shrinking, cracking, swelling, dissolution or change of mechanical qualities.

2.5 Tests of materials for cover, tapes and seams

The materials used for the cover, tapes, seams and additional equipment should be tested to the satisfaction of the Competent Authority to establish that they are rot-proof, colourfast and resistant to deterioration from exposure to sunlight and that they are not unduly affected by seawater, oil or fungal attack.

2.6 Strength tests

Body or lifting loop strength tests

2.6.1 The lifejacket should be immersed in water for a period of 2 min. It should then be removed from the water and closed in the same manner as when it is worn by a person. A force of not less than 3200 N (2400 N in the case of a child-size lifejacket) should be applied for 30 min to the part of the lifejacket that secures it to the body of the wearer (see figure 1) or to the lifting loop of the lifejacket. The lifejacket should not be damaged as a result of this test.

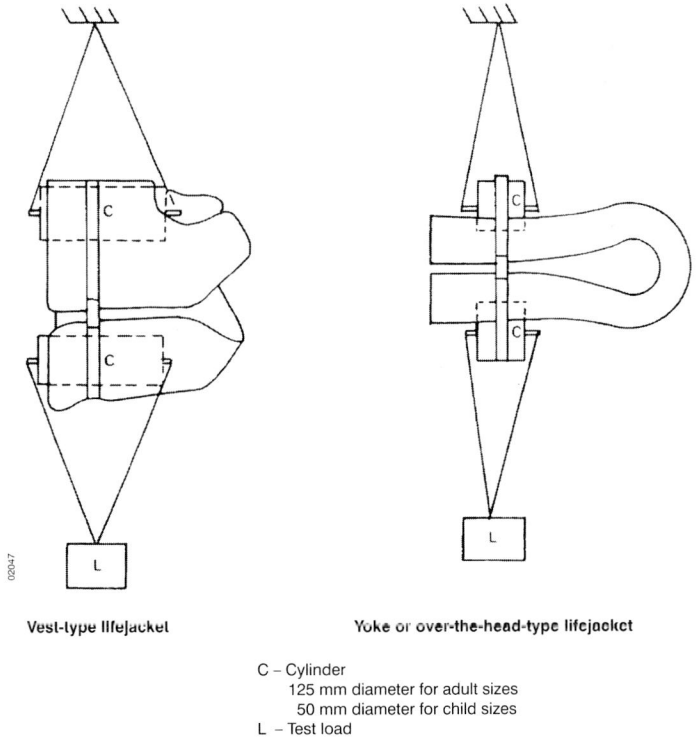

Vest-type lifejacket   Yoke or over-the-head-type lifejacket

C – Cylinder
125 mm diameter for adult sizes
50 mm diameter for child sizes
L – Test load

Figure 1 – Body strength test arrangement for lifejackets

Shoulder strength test

2.6.2 The lifejacket should be immersed in water for a period of 2 min. It should then be removed from the water and closed in the same manner as

Annex V

when it is worn by a person. A force of not less than 900 N (700 N in the case of a child-size lifejacket) should be applied for 30 min to the shoulder section of the lifejacket (see figure 2). The lifejacket should not be damaged as a result of this test.

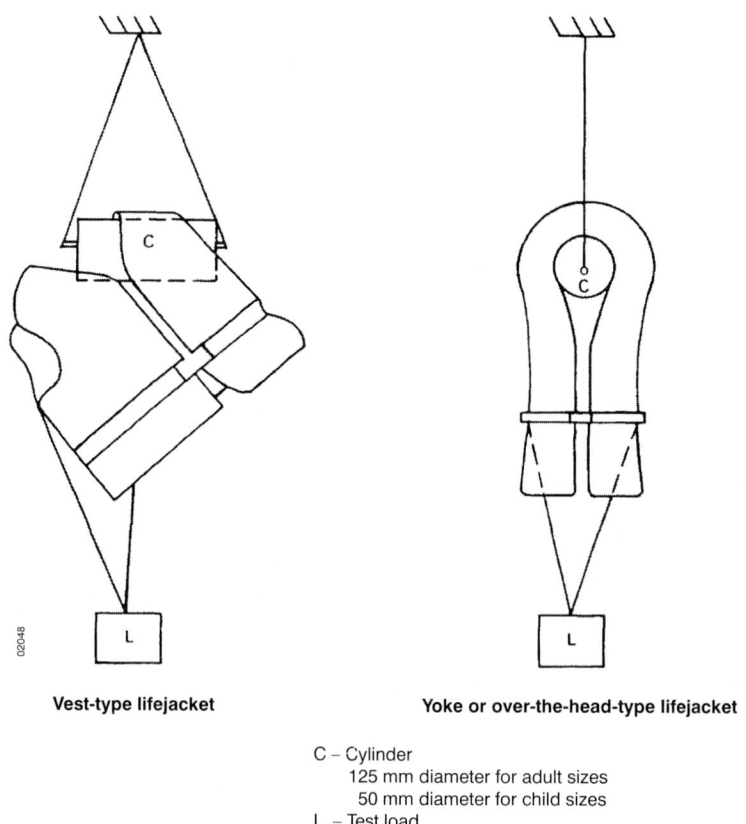

Vest-type lifejacket   Yoke or over-the-head-type lifejacket

C – Cylinder
125 mm diameter for adult sizes
50 mm diameter for child sizes
L – Test load

Figure 2 – Shoulder strength test arrangement for lifejackets

2.7 Additional tests for lifejacket buoyancy material other than cork or kapok

The following tests should be carried out on eight specimens of lifejacket buoyancy materials other than cork or kapok.

Test for stability under temperature cycling

2.7.1 Six specimens should be alternately subjected for 8 h to surrounding temperatures of –30°C and +65°C. These alternating cycles need not follow immediately after each other and the following procedure, repeated for ten cycles, is acceptable:

.1 an 8 h cycle at +65°C to be completed in one day;

.2 the specimens removed from the warm chamber that same day and left exposed under ordinary room conditions until the next day;

.3 an 8 h cycle at −30°C to be completed the next day; and

.4 the specimens removed from the cold chamber that same day and left exposed under ordinary room conditions until the next day.

2.7.2 The dimensions of the specimens should be recorded at the end of the 10-cycle period. The specimens should be carefully examined and should not show any sign of external change of structure or of mechanical qualities.

2.7.3 Two of the specimens should be cut open and should not show any sign of internal change of structure.

2.7.4 Four of the specimens should be used for water absorption tests, two of which should be so tested after they have also been subjected to the diesel oil test as prescribed in 1.2.

Tests for water absorption

2.7.5 The tests should be carried out in fresh water and the specimens should be immersed for a period of seven days under a 1.25 m head of water.

2.7.6 The tests should be carried out:

.1 on two specimens as supplied;

.2 on two specimens which have been subjected to the temperature cycling as prescribed in 2.7.1; and

.3 on two specimens which have been subjected to the temperature cycling as prescribed in 2.7.1 followed by the diesel oil test as prescribed in 2.4.

2.7.7 The specimens should be at least 300 mm square and be of the same thickness as used in the lifejacket. Alternatively, the entire lifejacket may be subjected to the test. The dimensions should be recorded at the beginning and end of these tests.

2.7.8 The results should state the mass in kilograms which each specimen could support out of the water after one and seven days immersion (the selection of a test method suitable for obtaining this result directly or indirectly is left to the discretion of the testing authority). The reduction of buoyancy should not exceed 16% for specimens which have been exposed to the diesel oil conditioning and should not exceed 5% for all other specimens. The specimens should show no sign of damage such as shrinking, cracking, swelling, dissolution or change of mechanical qualities.

2.8 Donning test

2.8.1 As lifejackets will be used by uninitiated persons, often in adverse conditions, it is essential that risk of incorrect donning be minimized. Ties and fastenings necessary for proper performance should be few and simple. Lifejackets should readily fit various sizes of adults, both lightly and heavily clad. Lifejackets should be capable of being worn inside-out, or clearly in only one way.

Test subjects

2.8.2 These tests should be carried out with at least six able-bodied persons of the following heights and weights:

Height	Weight
1.4 m–1.6 m	one person under 60 kg
	one person over 60 kg
1.6 m–1.8 m	one person under 70 kg
	one person over 70 kg
over 1.8 m	one person under 80 kg
	one person over 80 kg

   .1   at least one and not more than two of the persons should be females with not more than one female in the same height range; and

   .2   for the approval of the lifejackets, the test results obtained from each of the participating subjects should be acceptable except as provided otherwise.

Clothing

2.8.3 Each test subject should be tested wearing normal clothing. The test should be repeated with the test subject wearing heavy-weather clothing.

2.8.4 After demonstration, the test subjects should correctly don lifejackets within a period of 1 min, without assistance.

Assessment

2.8.5 The observer should note:

   .1   ease and speed of donning; and

   .2   proper fit and adjustment.

2.9 Water performance tests

2.9.1 This portion of the test is intended to determine the ability of the lifejacket to assist a helpless person or one in an exhausted or unconscious state and to show that the lifejacket does not unduly restrict movement. All tests should be carried out in fresh water under still conditions.

Test subjects

2.9.2 These tests should be carried out with at least six persons as described in 2.8.2. Only good swimmers should be used, since the ability to relax in the water is rarely otherwise obtained.

Clothing

2.9.3 Subjects should wear only swimming costumes.

Preparation for water performance tests

2.9.4 The test subjects should be made familiar with each of the tests set out below, particularly the requirement regarding relaxing and exhaling in the face-down position. The test subject should don the lifejacket unassisted using only the instructions provided by the manufacturer. The observer should note the points prescribed in 2.8.5.

Righting tests

2.9.5 The test subject should swim at least three gentle strokes (breast stroke) and then with minimum headway relax, with the head down and the lungs partially filled, simulating a state of utter exhaustion. The period of time should be recorded starting from the completion of the last stroke until the mouth of the test subject comes clear of the water. The above test should be repeated after the test subject has exhaled. The time should again be ascertained as above. The freeboard from the water surface to the mouth should be recorded with the test subject at rest.

Drop test

2.9.6 Without readjusting the lifejacket, the test subject should jump vertically into the water, feet first, from a height of at least 4.5 m. When jumping into the water, the test subject should be allowed to hold on to the lifejacket during water entry to avoid possible injury. The freeboard to the mouth should be recorded after the test subject comes to rest.

Assessment

2.9.7 After each of the water tests described above, the test subject should come to rest with the mouth clear of the water by at least 120 mm. The average of all subjects' trunk angles should be at least 30° back of vertical, and each individual subject's angle should be at least 20° back of vertical. The average of all subjects' faceplane (head) angles should be at least 40°

above horizontal, and each individual subject's angle should be at least 30° above horizontal. In the righting test, the mouth should be clear of the water in not more than 5 s. The lifejacket should not become dislodged or cause harm to the test subject.

2.9.8 When evaluating the results of a test in accordance with 2.9.5, 2.9.7 and 2.9.8, the Competent Authority may, in exceptional circumstances, disregard the results of a test on a subject if the results show a very slight deviation from the specified criteria, provided the Competent Authority is satisfied that the deviation can be attributed to the unusual size and stature characteristics of the test subject and the results of tests on other subjects, chosen in accordance with 2.9.2, show satisfactory performance of the lifejacket.

Swimming and water emergence test

2.9.9 All test subjects, without wearing the lifejacket, should attempt to swim 25 m and board a liferaft or a rigid platform with its surface 300 mm above the water surface. All test subjects who successfully complete this task should perform it again wearing the lifejacket. At least two thirds of the test subjects who can accomplish the task without the lifejacket should also be able to perform it with the lifejacket.

2.10 Children's lifejacket tests

As far as possible, similar tests should be applied for approval of lifejackets suitable for children.

2.10.1 When conducting water performance tests under 2.9, child-size lifejackets should meet the following requirements for their critical flotation stability characteristics. The range of sizes for child-size lifejackets should be considered, based on the test results. Devices should be sized by height or by height and weight.

2.10.2 Test subjects should be selected to fully represent the range of sizes for which the device is to be approved. Devices for smaller children should be tested on children as small as approximately 760 mm tall and 9 kg mass. At least six test subjects should be used for each 380 mm and 16 kg of size range:

    .1 *Turning time.* Each individual subject should turn face-up in not more than 5 s.

    .2 *Freeboard.* The combined results for clearance of the mouth above the water for all subjects should average at least 90 mm; each individual subject under 1270 mm and 23 kg should have at least 50 mm clearance, and each individual subject over 1270 mm and 23 kg should have at least 75 mm clearance.

    .3 *Trunk angle.* The average of all subjects' results should be at least 40° back of vertical, and each individual subject's result should be at least 20° back of vertical.

.4 *Faceplane (head) angle.* The average of all subjects' results should be at least 35° above horizontal, and each individual subject's result should be at least 20° above horizontal.

.5 *Mobility.* Mobility of the subject both in and out of the water should be given consideration in determining the acceptability of a device for approval.

2.11 Tests for inflatable lifejackets

2.11.1 Two inflatable lifejackets should be subjected to the temperature cycling test prescribed in 1.1 in the uninflated condition and should then be externally examined. The inflatable lifejacket materials should show no sign of damage such as shrinking, cracking, swelling, dissolution or change of mechanical qualities. The automatic and manual inflation systems should each be tested immediately after each temperature cycling test as follows:

.1 after the high-temperature cycle (test in 1.1.1), of two inflatable lifejackets taken from a stowage temperature of +65°C, one should be activated using the automatic inflation system by placing it in seawater at a temperature of +30°C and the other should be activated using the manual inflation system; and

.2 after the low-temperature cycle (test in 1.1.3), of two inflatable lifejackets taken from a stowage temperature of −30°C, one should be activated using the automatic inflation system by placing it in seawater at a temperature of −1°C and the other should be activated using the manual inflation system.

2.11.2 The test in 2.8 should be conducted using lifejackets both in the inflated and uninflated conditions.

2.11.3 The tests in 2.9 should be conducted using lifejackets that have been inflated both automatically and manually, and also with one of the compartments uninflated. The tests with one of the compartments uninflated should be repeated as many times as necessary to perform the test once with each compartment in the uninflated condition.

Tests of materials for inflatable bladders, inflation systems and components

2.11.4 The material used for the inflatable bladder, inflation system and components should be tested to establish that they are rot-proof, colour-fast and resistant to deterioration from exposure to sunlight and that they are not duly affected by seawater, oil or fungal attack.

Annex V

Material tests

2.11.5 Resistance to rot and illumination tested according to AATCC Method 30:1981 and ISO 105-B04:1988 Illumination should take place to class 4–5.

2.11.6 Following exposure to rot or illumination tests above, the tensile strength should be measured using the grab method given in ISO 5082. Minimum tensile strength should be not less than 300 N per 25 mm in the warp and weft direction.

Coated fabrics

2.11.7 Coated fabrics used in the construction of inflatable buoyancy chambers should comply with the following requirements:

.1 coating adhesion should be tested in accordance with ISO 2411:1991 by dropping the lifejacket from a height of 18 m into the water at 100 mm/min and should be not less than 50 N per 50 mm width;

.2 coating adhesion should be tested when wet following ageing according to ISO 188 with an exposure of 336 ± 0.5 h in fresh water at (70 ± 1)°C, following which the method at ISO 2411:1991 of dropping the lifejacket from a height of 18 m into the water at 100 mm/min and should not be less than 40 N per 50 mm width;

.3 tear strength should be tested in accordance with ISO 4674:1977, using method A1, and should not be less than 35 N;

.4 resistance to flex cracking should be tested in accordance with ISO 7854:1984 method A, using 9000 flex cycles, there should be no visible cracking or deterioration;

.5 breaking strength should be tested in accordance with ISO 1421:1977, using the CRE or CRT method, following conditioning for 24 ± 0.5 h at room temperature, and should not be less than 200 N per 50 mm width;

.6 breaking strength should be tested in accordance with ISO 1421:1977, using the CRE or CRT method, following conditioning immersed in fresh water for 24 ± 0.5 h at room temperature, and should not be less than 200 N per 50 mm width;

.7 elongation to break should be tested in accordance with ISO 1421:1977, using the CRE or CRT method, following conditioning at room temperature for 24 ± 0.5 h, and should not be more than 60%;

.8 elongation to break should be tested in accordance with ISO 1421:1977, using the CRE or CRT method, following conditioning immersed in fresh water at room temperature for 24 ± 0.5 h, and should not be more than 60%;

.9 the resistance to exposure to light when tested in accordance with ISO 105-B02:1988 and the contrast between the unexposed and exposed samples should not be less than class 5;

.10 the resistance to wet and dry rubbing when tested in accordance with ISO 105-X12:1995 and should not be less than class 3; and

.11 the resistance to seawater should not be less than class 4 in accordance with ISO 105-E02:1978 and the change in colour of the specimen should not be less than class 4.

Operating head load test

2.11.8 The operating head load test should be carried out using two lifejackets, one lifejacket to be conditioned at −30°C for 8 h and the other at +65°C for 8 h. After mounting on the manikin or the test form, the lifejackets should be inflated, and a steady force of 220 ± 10 N applied to the operating head as near as possible to the point where it enters the buoyancy chamber. This load should be maintained for 5 min, during which the direction and angle in which it is applied should be continuously varied. On completion of the test, the lifejacket should remain intact and should hold its pressure for 30 min.

Pressure test

2.11.9 The inflatable buoyancy chambers should be capable of withstanding an internal overpressure at ambient temperature. A lifejacket should be inflated using the manual method of inflation; after inflation, the relief valves should be disabled and a fully charged gas cylinder according to the manufacturer's recommendation should be fitted to the same inflation device and fired. The lifejacket should remain intact and should hold its pressure for 30 min. The lifejackets should show no signs of damage such as cracking, swelling or changes of mechanical qualities and that there has been no significant damage to the lifejacket inflation component. All fully charged gas cylinders used in this test should be sized according to the markings on the lifejacket.

2.11.10 With one buoyancy chamber inflated, the operating head on the opposite buoyancy chamber should be fired manually, using a fully charged gas cylinder according to the manufacturer's recommendations. The operation of the relief valves should be noted to ensure that the excess pressure is relieved. The lifejacket should remain intact and should hold its pressure for 30 min. The lifejacket should show no sign of damage such as cracking, swelling or changes of mechanical qualities and that there has been no significant damage to the lifejacket inflation component.

Air retention test

2.11.11 One inflation chamber of a lifejacket is filled with air until air escapes from the over-pressure valve or, if the lifejacket does not have an over-pressure valve, until its design pressure, as stated in the plans and specifications, is reached. After 12 h, the drop in pressure should not be greater than 10%. This test is then repeated as many times as necessary to test a different chamber until each chamber has been tested in this manner.

Compression test

2.11.12 The inflatable lifejacket, packed in the normal manner, should be laid on a table. A bag containing 75 kg of sand and having a base of 320 mm diameter should be lowered onto the lifejacket from a height of 150 mm in a time of 1 s. This should be repeated ten times, after which the bag should remain on the jacket for not less than 3 h. The lifejacket should be inflated by immersion into water and should inflate fully. The jacket should be inspected to ensure that no swelling or change of mechanical properties has occurred and checked for leaks.

Test of metallic components

2.11.13 Metal parts and components of a lifejacket should be corrosion-resistant to seawater and should be tested in accordance with ISO 9227:1990 for a period of 96 h. The metal components should be inspected and should not be significantly affected by corrosion or affect any other part of the lifejacket and should not impair the performance of the lifejacket.

2.11.14 Metal components should not affect a magnetic compass of a type used in small boats by more than $1°$ when placed at a distance of 500 mm from it.

Inadvertent inflation test

2.11.15 The resistance of an automatic inflation device to inadvertent operation should be assessed by exposing the entire lifejacket to sprays of water for fixed period. The lifejacket should be fitted correctly to a free-standing manikin of adult size, with a minimum shoulder height of 1,500 mm. The lifejacket should be deployed in the mode in which it is worn ready for use but not deployed as used in the water (i.e., if it is equipped with a cover which is normally worn closed, then the cover should be closed for the test). Two sprays should be installed so as to spray fresh water onto the lifejacket, as shown in the diagram. One should be positioned 500 mm above the highest point of the lifejacket, and at an angle of $15°$ from the vertical centreline of the manikin and the bottom line of the lifejacket. The other nozzle should be installed horizontally at a distance of 500 mm from the bottom line of the lifejacket, and pointed directly at the lifejacket. These nozzles should have a spray cone of $30°$, each orifice being 1.5 ± 0.1 mm in diameter, and the total area of the orifice should be 50 ± 5 mm^2, the orifice being evenly spread over the spray nozzle area.

2.11.16 The air temperature should be 20°C, and water should be supplied to the sprays at a pressure of 0.3 kPa to 0.4 kPa, a flow of 600 l/h, and a temperature of 18°C to 20°C.

2.11.17 The sprays should be turned on, and the lifejacket exposed to the following series of test to access the ability of the jacket to resist inadvertent inflation:

.1   5 min with the high spray on the front of the lifejacket;

.2   5 min with the high spray on the left side of the lifejacket;

.3   5 min with the high spray on the back of the lifejacket; and

.4   5 min with the high spray on the right side of the lifejacket.

2.11.18 During exposures specified in 2.11.17.1, 2.11.17.2 and 2.11.17.4 above, the horizontal spray should be applied for 10 periods of 3 s each to the front, left or right sides (but not back) as with the high spray.

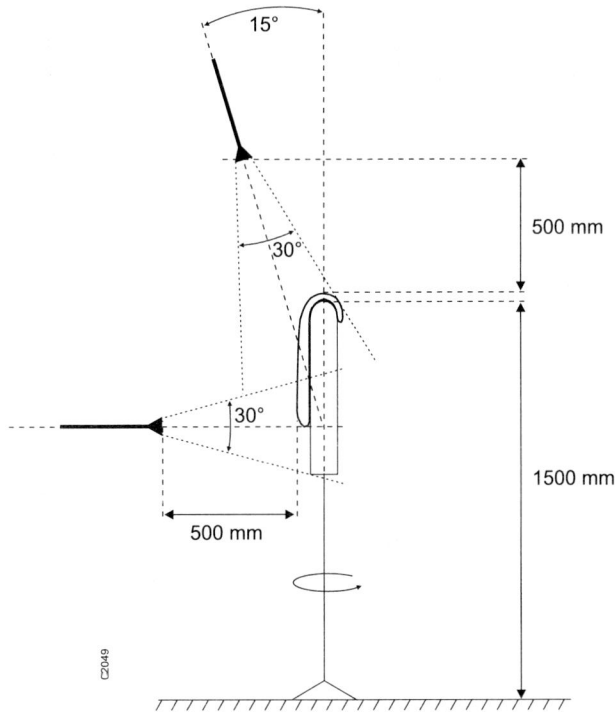

Figure 3 - *Test set-up for test of automatic inflation system*

Annex V

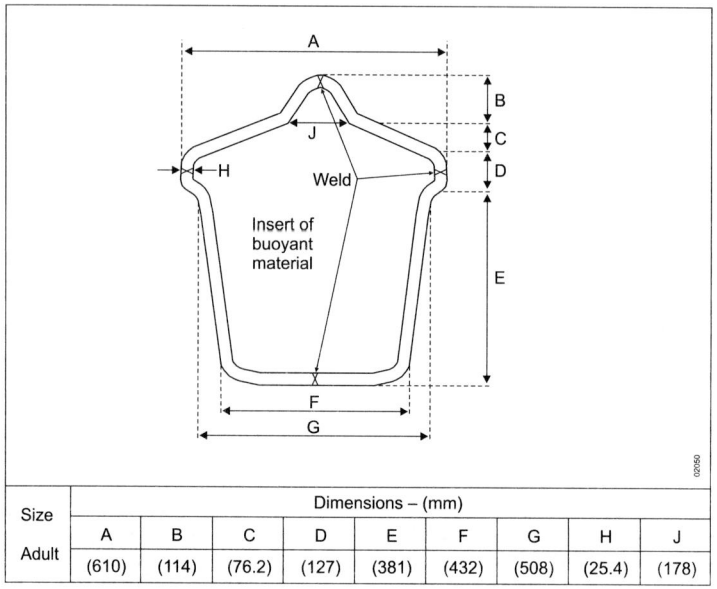

Size	Dimensions – (mm)							
	A	B	C	D	E	F	G	H	J
Adult	(610)	(114)	(76.2)	(127)	(381)	(432)	(508)	(25.4)	(178)

Figure 4 – *Alternative former*

2.11.19 After completing the above test, the lifejacket should be removed from the manikin and immersed in water to verify that the auto-inflation system functions.

Part 2
Production and installation tests

1   General

1.1   Representatives of the Competent Authority should make random inspection of manufacturers to ensure that the quality of life-saving appliances and the materials used comply with the specification of the approved prototype life-saving appliance.

1.2   Manufacturers should be required to institute a quality control procedure to ensure that life-saving appliances are produced to the same standard as the prototype life-saving appliance approved by the Competent Authority and to keep records of any production tests carried out in accordance with the Competent Authority instructions.

1.3   Where the proper operation of life-saving appliances is dependent on their correct installation in ships, the Competent Authority should require installation tests to ensure that the appliances have been correctly fitted in a ship.

2 Individual buoyancy equipment

Lifejackets

Production tests

2.1 Manufacturers should be required to carry out a buoyancy test on at least 0.5% of each batch of lifejackets produced, subject to a minimum of one from every batch.

Inspections by the Competent Authority

2.2 Inspections by a representative of the Competent Authority should be made at intervals of at least one per 6,000 lifejackets produced, subject to a minimum of one inspection per calendar quarter. When the manufacturer's quality control programme results in lifejackets that are consistently free of defects, the rate of inspection may be reduced to one in every 12,000. At least one lifejacket of each type in production should be selected at random by the inspector and subjected to detailed examination, including, if necessary, cutting open. The inspector should also be satisfied that the flotation tests are being conducted satisfactorily; if the inspector is not satisfied, a flotation test should be undertaken.

Annex VI

Recommended standards for pilot ladders
(Regulation 23 of SOLAS chapter V:
Pilot transfer arrangements)

1   Application

1.1   Ships engaged on voyages in the course of which pilots are likely to be employed shall be provided with pilot transfer arrangements.

1.2   Equipment and arrangements for pilot transfer which are installed on or after 1 January 1994 shall comply with the requirements of this regulation, and due regard shall be paid to the standards adopted by the Organization.*

1.3   Equipment and arrangements for pilot transfer which are provided on ships before 1 January 1994 shall at least comply with the requirements of regulation 17 of the International Convention for the Safety of Life at Sea, 1974 in force prior to that date, and due regard shall be paid to the standards adopted by the Organization prior to that date.

1.4   Equipment and arrangements which are replaced after 1 January 1994 shall, in so far as is reasonable and practicable, comply with the requirements of this regulation.

2   General

2.1   All arrangements used for pilot transfer shall efficiently fulfil their purpose of enabling pilots to embark and disembark safely. The appliances shall be kept clean, properly maintained and stowed and shall be regularly inspected to ensure that they are safe to use. They shall be used solely for the embarkation and disembarkation of personnel.

2.2   The rigging of the pilot transfer arrangements and the embarkation of a pilot shall be supervised by a responsible officer having means of

---

* Refer to the Recommendation on pilot transfer arrangements, adopted by the Organization by resolution A.889(21), and to MSC/Circ.568/Rev.1 on Required boarding arrangements for pilots.

communication with the navigation bridge who shall also arrange for the escort of the pilot by a safe route to and from the navigation bridge. Personnel engaged in rigging and operating any mechanical equipment shall be instructed in the safe procedures to be adopted and the equipment shall be tested prior to use.

3 Transfer arrangements

3.1 Arrangements shall be provided to enable the pilot to embark and disembark safely on either side of the ship.

3.2 In all ships, where the distance from sea level to the point of access to, or egress from, the ship exceeds 9 m, and when it is intended to embark and disembark pilots by means of the accommodation ladder, or by means of mechanical pilot hoists or other equally safe and convenient means in conjunction with a pilot ladder, the ship shall carry such equipment on each side, unless the equipment is capable of being transferred for use on either side.

3.3 Safe and convenient access to, and egress from, the ship shall be provided by either:

.1 a pilot ladder requiring a climb of not less than 1.5 m and not more than 9 m above the surface of the water, so positioned and secured that:

.1.1 it is clear of any possible discharges from the ship;

.1.2 it is within the parallel body length of the ship and, as far as is practicable, within the mid-ship half length of the ship;

.1.3 each step rests firmly against the ship's side; where constructional features, such as rubbing bands, would prevent the implementation of this provision, special arrangements shall, to the satisfaction of the Administration, be made to ensure that persons are able to embark and disembark safely;

.1.4 the single length of pilot ladder is capable of reaching the water from the point of access to, or egress from, the ship and due allowance is made for all conditions of loading and trim of the ship, and for an adverse list of 15°; the securing strong point, shackles and securing ropes shall be at least as strong as the side ropes;

.2 an accommodation ladder in conjunction with the pilot ladder, or other equally safe and convenient means, whenever the distance from the surface of the water to the point of access to the ship is more than 9 m. The accommodation ladder shall be sited leading aft. When in use, the lower end of the accommodation ladder shall rest firmly against the ship's side within the parallel body length of the ship and, as far as is practicable, within the mid-ship half length and clear of all discharges; or

Annex VI

.3 a mechanical pilot hoist so located that it is within the parallel body length of the ship and, as far as is practicable, within the mid-ship half length of the ship and clear of all discharges.

4 Access to the ship's deck

Means shall be provided to ensure safe, convenient and unobstructed passage for any person embarking on, or disembarking from, the ship between the head of the pilot ladder, or of any accommodation ladder or other appliance, and the ship's deck. Where such passage is by means of:

.1 a gateway in the rails or bulwark, adequate handholds shall be provided;

.2 a bulwark ladder, two handhold stanchions rigidly secured to the ship's structure at or near their bases and at higher points shall be fitted. The bulwark ladder shall be securely attached to the ship to prevent overturning.

5 Shipside doors

Shipside doors used for pilot transfer shall not open outwards.

6 Mechanical pilot hoists

6.1 The mechanical pilot hoist and its ancillary equipment shall be of a type approved by the Administration. The pilot hoist shall be designed to operate as a moving ladder to lift and lower one person on the side of the ship, or as a platform to lift and lower one or more persons on the side of the ship. It shall be of such design and construction as to ensure that the pilot can be embarked and disembarked in a safe manner, including a safe access from the hoist to the deck and vice versa. Such access shall be gained directly by a platform securely guarded by handrails.

6.2 Efficient hand gear shall be provided to lower or recover the person or persons carried, and kept ready for use in the event of power failure.

6.3 The hoist shall be securely attached to the structure of the ship. Attachment shall not be solely by means of the ship's side rails. Proper and strong attachment points shall be provided for hoists of the portable type on each side of the ship.

6.4 If belting is fitted in the way of the hoist position, such belting shall be cut back sufficiently to allow the hoist to operate against the ship's side.

6.5 A pilot ladder shall be rigged adjacent to the hoist and be available for immediate use so that access to it is available from the hoist at any point of its travel. The pilot ladder shall be capable of reaching sea level from its own point of access to the ship.

6.6   The position on the ship's side where the hoist will be lowered shall be indicated.

6.7   An adequate protected stowage position shall be provided for the portable hoist. In very cold weather, to avoid the danger of ice formation, the portable hoist shall not be rigged until its use is imminent.

7   Associated equipment

7.1   The following associated equipment shall be kept at hand ready for immediate use when persons are being transferred:

   .1   two man-ropes of not less than 28 mm in diameter, properly secured to the ship, if required by the pilot;

   .2   a lifebuoy equipped with a self-igniting light; and

   .3   a heaving line.

7.2   When required by 4, stanchions and bulwark ladders shall be provided.

8   Lighting

Adequate lighting shall be provided to illuminate the transfer arrangements overside, the position on deck where a person embarks or disembarks and the controls of the mechanical pilot hoist.

Annex VII
Annotated list of pertinent publications

FAO (www.fao.org)

FAO Code of Conduct for Responsible Fisheries
The Code sets out principles and international standards of behaviour for responsible practices with a view to ensuring the effective conservation, management and development of living aquatic resources, with due respect for the ecosystem and biodiversity.

FAO Technical Guidelines for Responsible Fisheries – Fishing Operations
The technical guidelines are given in support of the implementation of the Code of Conduct in relation to fishing operations. They are addressed to States, international organizations, fisheries management bodies, owners, managers and charterers of fishing vessels, and fishermen and their organizations.

FAO Standard Specifications for the Marking and Identification of Fishing Vessels
This document contains the specifications of a standardized system for the marking and identification of fishing vessels as endorsed by the FAO Committee on Fisheries, Rome, April 1989.

FAO Safety at sea as an integral part of fisheries management
This paper provides a comprehensive overview of sea safety issues, and concludes that safety at sea should be integrated into fisheries management.

IMO (www.imo.org)

Convention on the International Regulations for Preventing Collisions at Sea, 1972 (COLREG)
The Convention on the International Regulations for Preventing Collisions at Sea, 1972 has been accepted by many States since it was adopted in 1972 and entered into force in July 1977. It was amended in 1981, 1987,

1989, 1993 and 2001. The publication contains the fully consolidated text of the 1972 Convention.

The 1993 Torremolinos Protocol and Torremolinos International Convention for the Safety of Fishing Vessels (Consolidated edition, 1995)
The publication contains consolidated text of the regulations annexed to the Torremolinos International Convention for the Safety of Fishing Vessels, 1977, as modified by the Torremolinos Protocol of 1993 relating thereto.

Code on Intact Stability for All Types of Ships Covered by IMO Instruments (resolution A.749(18), as amended)
The Code has been assembled to provide, in a single document, recommended provisions relating to intact stability, based primarily on existing IMO instruments.

Code of Practice concerning the Accuracy of Stability Information for Fishing Vessels (resolution A.267(VIII))

Recommended Practice on Portable Fish-Hold Divisions (resolution A.168(ES.IV), appendix V, as amended by resolution A.268(VIII))

Fire Test Procedures Code (resolution MSC.61(67))

Fire Safety Systems Code (resolution MSC.98(73))

Recommendation on improved fire test procedures for flammability of bulkheads, ceiling and deck finish materials (resolution A.653(16))

Guidelines on the evaluation of fire hazard properties of materials (resolution A.166(ES.IV))

Improved guidelines for marine portable fire extinguishers (resolution A.951(23))

Graphical symbols for shipboard fire control plans (resolution A.952(23))

Recommendation on fire test procedures for ignitability of primary deck coverings (resolution A.687(17))

Life-Saving Appliances Code (LSA Code) (resolution MSC.48(66))

Revised recommendations on the testing of life-saving appliances (resolution MSC.81(70), as revised)

Code of Practice for the evaluation, testing and acceptance of prototype novel life-saving appliances and arrangements (resolution A.520(13))

Standardized life-saving appliance evaluation and test report forms (MSC/Circ.980)

Recommendation on performance standards for magnetic compasses (resolution A.382(X), annex I)

Annex VII

Recommendation on performance standards for gyro-compasses (resolution A.424(XI))

Recommendation on performance standards for radar equipment (resolution MSC.64(67), annex 4)

Performance standards for automatic radar plotting aids (ARPAs) (resolution A.823(19))

Performance standards for survival craft radar transponders for use in search and rescue operations (resolution A.802(19))

Recommendation on performance standards for echo-sounding equipment (resolution A.224(VII), as amended by resolution MSC.74(69), annex 4)

Recommendation on performance standards for devices to indicate speed and distance (resolution A.824(19), as amended by resolution MSC.96(72))

Performance standards for rate-of-turn indicators (resolution A.526(13))

Recommendation on unification of performance standards for navigational equipment (resolution A.575(14))

Recommendation on methods of measuring noise levels at listening posts (resolution A.343(IX))

Recommendation on performance standards for shipborne global positioning system receiver equipment (resolution A.819(19), as amended by resolution MSC.112(73))

Recommendation on performance standards for shipborne GLONASS receiver equipment (resolution MSC.53(66), as amended by resolution MSC.113(73))

Recommendation on performance standards for combined GPS/GLONASS receiver equipment (resolution MSC.74(69), annex 1, as amended by resolution MSC.115(73))

Recommendation on performance standards for heading control systems (resolution MSC.64(67), annex 3)

Recommendation on performance standards for shipborne Loran-C and Chayka receivers (resolution A.818(19))

Recommendation on performance standards for shipborne DGPS and DGLONASS maritime radio beacon receiver equipment (resolution MSC.64(67), annex 2, as amended by resolution MSC.114(73))

Recommendation on performance standards for track control systems (resolution MSC.74(69), annex 2)

Recommendation on performance standards for a universal shipborne automatic identification system (AIS) (resolution MSC.74(69), annex 3)

Recommendation on performance standards for radar reflectors (resolution A.384(X), as amended by resolution MSC.164(78))

Recommendation on performance standards for sound reception systems (resolution MSC.86(70), annex 1)

Recommendation on performance standards for voyage data recorders (VDRs) (resolution A.861(20))

Recommendation on performance standards for electronic chart display and information systems (ECDIS) (resolution A.817(19)), as amended by resolutions MSC.64(67), annex 5, and MSC.86(70), annex 4)

Recommendation on performance standards for daylight signalling lamps (resolution MSC.95(72))

NAVTEX Manual (publication IC951E)

Provision of radio services for the global maritime distress and safety system (GMDSS), (resolution A.704(17))

Carriage of radar operating in the frequency band 9,300–9,500 MHz (resolution A.614(15))

Carriage of Inmarsat Enhanced Group Call SafetyNET receivers under the global maritime distress and safety system (GMDSS) (resolution A.701(17))

Promulgation of maritime safety information (resolution A.705(17))

Search and rescue homing capability (resolution A.616(15))

Operational standards for radiotelephone alarm signal generators (resolution A.421(XI))

Performance standards for narrow-band direct-printing telegraph equipment for the reception of navigational and meteorological warnings and urgent information to ships (resolution A.525(13))

General requirements for shipborne radio equipment forming part of the global maritime distress and safety system (GMDSS) and for electronic navigational aids (resolution A.694(17))

Performance standards for ship earth stations capable of two-way communications (resolution A.698(17))

Type approval of ship earth stations (resolution A.570(14))

Performance standards for shipborne VHF radio installations capable of voice communication and digital selective calling (resolution A.609(15))

Performance standards for shipborne MF radio installations capable of voice communication and digital selective calling (resolution A.610(15))

Performance standards for shipborne MF/HF radio installations capable of voice communication, narrow-band direct printing and digital selective calling (resolution A.613(15))

Performance standards for float-free satellite emergency position-indicating radio beacons (EPIRBs) operating on 406 MHz (resolution A.695(17))

Annex VII

Type approval of satellite emergency position-indicating radio beacons (EPIRBs) operating in the COSPAS-SARSAT system (resolution A.696(17))

Performance standards for survival craft radar transponders for use in search and rescue operations (resolution A.697(17))

Performance standards for float-free VHF emergency position-indicating radio beacons (resolution A.612(15))

Performance standards for Inmarsat standard-C ship earth stations capable of transmitting and receiving direct-printing communications (resolution A.663(16))

Performance standards for enhanced group call equipment (resolution A.664(16))

Performance standards for float-free satellite emergency position-indicating radio beacons operating through the geostationary Inmarsat satellite system on 1.6 GHz (resolution A.661(16))

Performance standards for float-free release and activation arrangements for emergency radio equipment (resolution A.662(16))

System performance standards for the promulgation and co-ordination of maritime safety information using high-frequency narrow-band direct printing (resolution A.699(17))

Performance standards for narrow-band direct-printing telegraph equipment for the reception of navigational and meteorological warnings and urgent information to ships (MSI) by HF (resolution A.700(17))

Code on Noise Levels on Board Ships (resolution A.468(XII))

Pilot transfer arrangements (SOLAS Convention, 1974, as amended, regulation 23 of chapter V)

WHO (www.who.int/en/org)

International Medical Guide for Ships
The ILO/IMO/WHO *International Medical Guide for Ships*, published by the World Health Organization, is intended for use by people, with little or no formal medical training, who are responsible for health care on board ships of all kinds.

Others

European Union Council Directive 92/29/EEC on minimum safety and health requirements for improved medical treatment on board vessels

IEC Publication 60079

Information note
Fisheries management measures

The Information Note to part B of the Code of Safety for Fishermen and Fishing Vessels of 1975 reflected elements of fisheries management governing the operation of fishing fleets in the major fishing regions of the world that were common at that time. However, although some of the vessels listed in the information note are still in operation, this approach to the management of fishing operations has greatly changed since the adoption of the United Nations Convention on the Law of the Sea of 1982.

By 1989, with the involvement of the International Maritime Organization (IMO), the Food and Agriculture Organization of the United Nations (FAO) had developed "The Standard Specifications for the Marking of Fishing Vessels"* for adoption by States on a voluntary basis as an aid to fisheries management and safety at sea. These standards were endorsed by the Eighteenth Session of the FAO Committee on Fisheries and IMO and the International Telecommunications Union were informed accordingly by the Director-General of FAO. Essentially, these standards require the International Radio Call Sign (IRCS) allocated to a fishing vessel to be prominently displayed on both sides of the vessel and on a horizontal surface. Provisions were also developed for vessels that have not been allocated an IRCS.

In more recent years, it has been acknowledged that fishing vessel safety is an integral part of fisheries management and this is reflected in the way in which fishing operations are managed by coastal States and monitored and managed by regional fisheries management organizations in areas outside national jurisdiction. In addition, the Code of Conduct for Responsible Fisheries, which was unanimously adopted on 31 October 1995 by the FAO Conference, contains the principle that States should make arrangements individually, together with other States, or with the appropriate international organization to integrate fishing operations into maritime search and rescue systems.

In general, vessels that are authorized to fish in waters under national jurisdiction are subject to a reporting regime (position and catch) and are often required to report by a vessel monitoring system (VMS). The VMS equipment on the fishing vessel periodically reports the positions of the fishing vessel and other data to the fisheries management authorities at

---

* Described in the FAO Technical Guidelines for Responsible Fisheries – No 1 Fishing Operations. (ISBN 92-5-103914-3) and MSC/Circ.572.

Fisheries management measures

national, regional or international levels. VMS is now widely used by fisheries authorities to monitor their own fleets and, in particular, to monitor those foreign flag vessels that have sought access to the resources of the coastal State, and this has been seen to greatly benefit conventional fisheries monitoring, control and surveillance measures.

Reporting by VMS is generally more frequent and interactive than the reporting by LRIT (Long-Range Identification and Tracking of ships) or by Vessel Reporting Systems. This is because passenger and cargo ships are transiting more predictable routes whereas fishing vessels tend to change course and speed much more in their fishing operations and their positions are less predictable. The pattern of positions of the fishing vessels displayed on a monitor can give the fisheries management authorities a great deal of information on the activities that a fishing vessel might or might not be undertaking.*

It should be noted, however, that the vessel equipment used in VMS may or may not be GMDSS-compliant. This is because there are other VMS service providers in addition to Inmarsat in satellite communications and some VMS operate one-way communications which do not allow polling. In the case that the satellite transmitters are not GMDSS-compliant, valuable information about a vessel in distress or information of other fishing vessels in the area might not be available to the Search and Rescue (SAR) services. For this reason, SAR authorities are encouraged to familiarise themselves with the national and regional availability of such systems and, if necessary, liaise with the fisheries management authorities who operate such systems, particularly since these fishing vessels should be participating to the extent required in any SAR operation.

On the high seas, regional fisheries management organizations are charged by the Contracting Parties to ensure the optimum utilization, rational management and conservation of the fishery resources of a regulatory area and to promote international co-operation. In general, each organization maintains a record of all fishing vessels that have been authorized to fish in the regulatory area by the Contracting Parties. In many of the regulatory areas, the fishing operations are monitored by on-board observers, by patrol vessels and by aircraft. In some cases, the actual patrol vessels may be a commercial fishing vessel provided by one of the Contracting Parties for a specific period of time during which it is not allowed to engage in fishing. Where on-board observers are carried, they are expected to report on fishing activities and discards, verify entries in log-books, report the vessel position, collect data on catch and effort, and collect samples for scientific work. In addition, several fisheries management organizations require that the VMS reports made by these vessels to their flag States, while the vessels are in the regulatory area, are transmitted to the fisheries management organization. This data can be

---

* Further information has been published in the FAO Technical Guidelines for Responsible Fisheries No.1 Suppl. 1 – Vessel Monitoring Systems. (ISBN 92-5-104179-2).

forwarded to fishery patrol vessels or aircraft in the regulatory area, in order that fishing vessels can be located and inspected. This information on the positions of fishing vessels in the area could supplement the information reported to the SAR by GMDSS and assist the fisheries patrol craft and fishing vessels meet their obligations in a distress situation if required or if requisitioned.* This is particularly beneficial in very remote areas far from shipping lanes where alternative sources of assistance might be absent. Further information on the network of FAO and non-FAO regional fisheries management bodies can be obtained at http://www.fao.org/fi/body/rfb/chooseman_type.htm.

Furthermore, as an indication of the increasing control exerted by flag States over their fishing vessels on the high seas, there is the Agreement to Promote Compliance with International Conservation and Management Measures by Fishing Vessels on the High Seas (FAO Compliance Agreement). This requires, *inter alia*, that only vessels so authorized by a Party are allowed to fish on the high seas. This Agreement also requires Parties to provide FAO with particulars of vessels so authorized. Thus, FAO maintains a database containing details of such vessels and attendant fisheries management information that can be shared by the Parties to the Agreement. The Compliance Agreement entered into force on 24 April 2003.

In addition, recent initiatives in satellite remote sensing have combined VMS positions with images obtained from satellites using Synthetic Aperture Radar. Comparing the radar images with the VMS data can automatically detect vessels on the image which are not reporting by VMS or reporting a false position and, therefore, likely to be involved in Illegal, Unregulated or Unreported Fishing (IUU Fishing). The EU Joint Research Centre at Ispra in Italy has been co-ordinating the project with the participation of the fisheries management authorities of most EU countries. The project also includes Norway, Iceland and Canada and the North East Atlantic Fisheries Commission (NEAFC) that has the mandate for fisheries management in the high seas area outside national jurisdictions in the region. The project has been very successful in remote areas, where the targets are very likely to be fishing vessels. However, for SAR operations it is necessary that SAR Authorities are aware of all vessels in the vicinity of a casualty and the type of information provided with satellite imagery would be extremely useful in a distress incident. Such information would be particularly useful in areas remote from shipping lanes where there are few vessels to assist a vessel in distress.

---

* Refer to SOLAS regulation V/33.

Fisheries management measures

System Overview

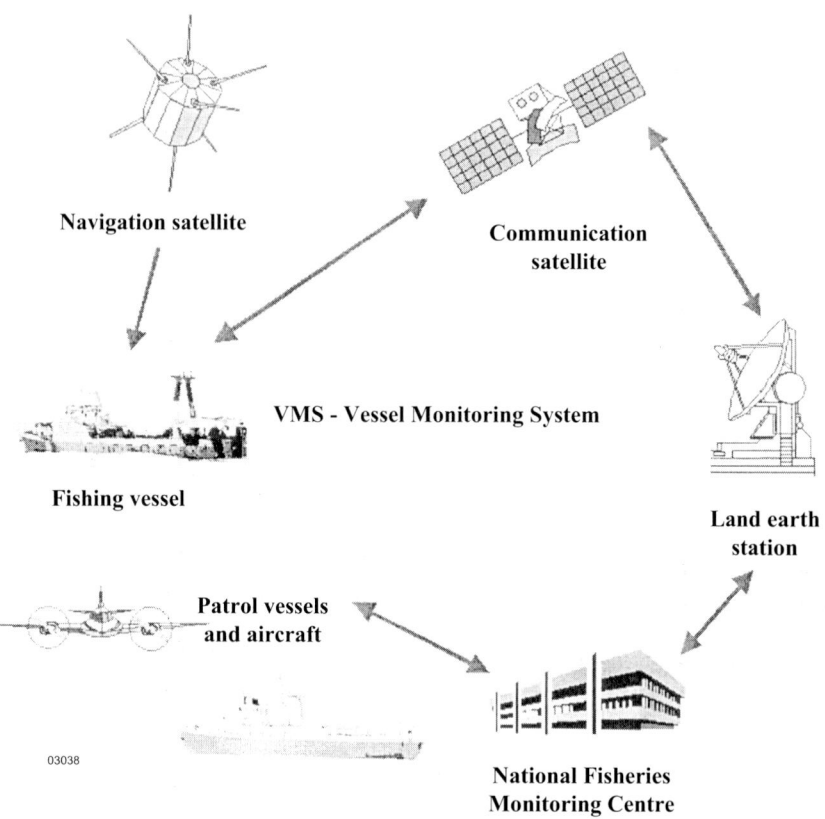

Index

abandon-ship training   121

category A   64

chilled seawater (CSW)   15

"A" class divisions   48

"B" class divisions   49

"C" class divisions   50

Code on Intact Stability for all Types of Ships Covered by IMO Instruments   19

distress signals   116

drills   121

DSC   127

emergency alarm system   120

EPIRBs   128

"F" class divisions   50

Fire control plan   75, 87, 98

fire extinguishers   72, 85, 96

fire hoses   70, 84, 96

fire hydrants   70, 84, 96

fire mains   70, 84, 95

fire pumps   69, 83, 94

Fire Safety Systems Code   50

Fire Test Procedures Code   50

firefighters' outfits   75, 87, 98

GZ   16

ice accretion   20

immersion suits   115

inclining test   21

international shore connection   74

lifebuoys   116

lifejackets   115

line-throwing appliances   116

maximum permissible operating draught   22

metacentric height GM_o 17

method IF   48

method IIF   48

method IIIF   48

muster list   120

protective measures   99

radar transponders   117

refrigerated seawater (RSW)   15

righting lever   17

sanitary facilities   149

stability information   21

subdivision and damage stability   23

survival craft   113

thermal protective aids   115

training   124

ventilation systems   58, 78, 90

watertight doors   8

weathertight doors   9

197

Notes